American Bison

Organisms and Environments

HARRY W. GREENE, CONSULTING EDITOR

1. *The View from Bald Hill: Thirty Years in an Arizona Grassland,* by Carl E. Bock and Jane H. Bock

2. *Tupai: A Field Study of Bornean Treeshrews,* by Louise H. Emmons

3. *Singing the Turtles to Sea: The Comcáac (Seri) Art and Science of Reptiles,* by Gary Paul Nabhan

4. *Amphibians and Reptiles of Baja California, Including Its Pacific Islands and the Islands in the Sea of Cortés,* by L. Lee Grismer

5. *Lizards: Windows to the Evolution of Diversity,* by Eric R. Pianka and Laurie J. Vitt

6. *American Bison: A Natural History,* by Dale F. Lott

American Bison

A NATURAL HISTORY

Dale F. Lott

WITH A FOREWORD BY HARRY W. GREENE

University of California Press   Berkeley   Los Angeles   London

University of California Press
Berkeley and Los Angeles, California

University of California Press, Ltd.
London, England

First paperback printing 2003

The photographs showing display hair, display hair loss, copulation,
and ejaculation originally appeared in *Zeitschrift für Tierpsychologie* 49
(1979) and 56 (1981) and are reprinted by the kind permission of
Blackwell Wissenschafts-Verlag Berlin, GmbH.

Library of Congress Cataloging-in-Publication Data

Lott, Dale F.
American bison : a natural history / Dale F. Lott ; with a foreword by
Harry W. Greene.
    p.   cm. — (Organisms and environments ; 6)
    Includes bibliographical references (p.   ) and index.
    ISBN 0-520-24062-6 (pbk : alk. paper)
    1. American bison. I. Title. II. Series.
    QL737.U53 L68   2002
    599.64'3—dc21   2002000243

Manufactured in the United States of America

10   09   08   07   06   05   04   03
10   9   8   7   6   5   4   3   2   1

The paper used in this publication is both acid-free and totally
chlorine-free (TCF). It meets the minimum requirements of
ANSI/ NISO Z39.48-1992 (R 1997) (*Permanence of Paper*). ∞

For my father

Contents

Foreword

Harry W. Greene

American Bison: A Natural History is the sixth volume in the University of California Press series on organisms and environments. Our unifying themes are the diversity of plants and animals, the ways in which they interact with each other and with their surroundings, and the broader implications of those relationships for science and society. We seek books that promote unusual, even unexpected connections among seemingly disparate topics, and we want to encourage projects that are distinguished by the unique perspectives and talents of their authors. Other volumes thus far have concerned the ecology of Arizona grasslands, the behavior of Bornean treeshrews, Seri ethnoherpetology, the amphibians and reptiles of Baja California, and global lizard biology.

Among all living mammals, bison are perhaps most emblematic of North America and of the fate of wildlife on a continent so long and dramatically transformed, especially under the weight of our impact. During their most recent twenty thousand years or so, bison have witnessed glaciers advancing and receding with widespread climate changes, invasions of the New World by humans from Asia and Europe, and the extinction of several dozen species of other large mammals. Over those same millennia we have slaughtered bison in numbers beyond comprehension, as men first with spears chased them over cliffs, then later with bow and arrows dispatched them from horseback, and finally with rifles shot them from railway cars. Today bison still roam North American prairies, most of them in a few small herds from Canada to Oklahoma, and their biology still mirrors the seasonal rhythms of weather and local ecology. And as did their Pleistocene ancestors, these massive, shaggy creatures stir up the insects that cowbirds eat, crop vast expanses of grassland, and create vernal pools with their dust wallows. Meanwhile, bison are threatened by ever-shrinking habitat, political controversies, and, perhaps surprisingly, the possibility of extinction through domestication.

Ask professional animal behaviorists what motivates their field studies and the commonest answer is likely "curiosity," especially a desire to know what it is like to be a particular animal. Although some biologists and philosophers regard that question as fundamentally unanswerable, many of us just as stubbornly find it irresistible, and there's no denying that accurate observations provide windows into the lives of other organisms. Thoroughly documented field biology also supplies the facts that underlie conservation, management, and public concern for nature. *American Bison* serves all of these goals as it follows one man's lifetime obsession with the details of bison natural history. Dale Lott is a respected scholar as well as a keen and sympathetic observer of animals and people, and I think he must love bison. Lott grew up among these creatures he often as not calls buffalo, and his engaging prose crackles, almost rumbles at times with a deep appreciation for them and their surroundings. Beyond the basic findings of science, he offers pithy comments on rural lifestyles and just enough personal background to give the reader a welcome familiarity and confidence in his views. Perhaps most intriguingly, Lott asks us to distinguish between real and mythical bison, between the animals as they were and still might be and the animals as we wish they were. With honest knowledge, he seems to imply, we might better ask what kind of bison we will tolerate and what kind of world we can share.

American Bison comprehensively summarizes the biology of these large mammals. Thanks to Lott's interest in seemingly everything having to do with the Great Plains, along the way we learn about mountains, rain shadows, wind, soil, wolves, prairie dogs, horses, symbiotic gut bacteria, and diseases of cattle. He also tutors us in the basics of ecological theory and behavioral endocrinology, as well as in what to do when a bison cow charges. This wonderful book is packed with facts, some familiar, others new and even surprising. For example, there are likely no more than 30 million bison at peak abundance in North America, rather than 60 or 70 million as is often stated. Ironically, all the founder animals of today's public herds—except for those in Yellowstone and the wood bison in Canada—came from the stock of famed Texas rancher Charles Goodnight and other private owners. Badgers and coyotes sometimes do hunt col-laboratively, as long claimed by Indians; and however awkward a grown bison might look to us, one bull did a standing high jump that landed him six feet above his launch site, and another "hopped" an eight-foot-wide cattle guard!

We are so fortunate that today there are places in North America to ex-perience such magnificent creatures and that Dale Lott's life's work in-spires us to learn what it's like to be one of them. In the parks and refuges where bison still live freely, a visitor can stand near a herd during mating season and feel the tremble of their movements and vocalizations, or watch a ton of woolly brown herbivore casually drop to the ground and roll in a cloud of dust. *American Bison: A Natural History* closes with care-fully reasoned arguments for a Great Plains Park, a place where we can preserve these animals and thereby save some wilderness for all of us.

Preface

I had the great good luck to live the first six years of my life in a place where buffalo were the animals I saw most often—the National Bison Range in western Montana. My mother's father was the Range superintendent; my father was a Range employee who had grown up near the Range and had married the boss's daughter. My older brother and I lived on the Range with them in staff housing.

When I was ten my parents bought my father's father's farm/ranch three miles from the Bison Range. Through my adolescence the bison herds were visible as dark patches on distant hills, though we occasionally went to the Range to see them up close. College, the army, a Ph.D. in psychology followed by a postdoc at the Institute for Animal Behavior in Rutgers—and I was a thirty-year-old, newly hired assistant professor at the University of California at Davis looking for a suitable subject for field research. Bison came to mind.

There were good scientific reasons. Their behavior was not yet thoroughly described and was scarcely analyzed at all. They are easy to see in the grasslands they frequent and are active during the day. Then too, I really liked them—always had. I stopped by the Bison Range on my way from New Jersey to California and began to collect data.

I first encountered bison not as symbols of the West, the squandering of a natural resource, or a conservation triumph. They were simply the animals I had seen most often when I was a young child—enthralling in and of themselves. I still see them that way, and my research goal was to know and understand them better. The basic question has always been, how do they get on in the world?

This is, to use a couple of very old-fashioned terms, a natural history of buffalo. After years of living near them, and still more years of doing field research on them, I'm still fascinated by bison and hope to share some of the magic of my experiences and the things I learned—things about bison and the ecological community they evolved to be a part of. Eventually, people made themselves a part of that community. Their role in it, and in the fate of the bison, must be told too.

Animals conduct transactions with their environment, and especially with the other members of their community, primarily through their behavior. Consequently, understanding behavior is fundamental to understanding both the animals and their interactions. I'm happy about that, because I find animal behavior in general fully as fascinating as I find bison behavior in particular.

It's hard to imagine—as you fly above mile after mile of corn, soybeans, and cattle feedlots or drive between them—but before East Africa became safari land, rich adventurers went on safari on the Great Plains. Buffalo Bill got his start in show business by laying on a safari for the Czarevitch of Russia—the Grand Duke Alexis. The Great Plains was a good choice. This vast, little-disturbed natural community covered a third of the United States—creating wonder, inviting adventure. Part of the appeal was the exotic indigenous people, but the main attraction was a sea of grass inhabited by an assemblage of animals mostly unknown elsewhere, and dominated by enormous herds of buffalo.

This community no longer exists anywhere, though nearly all the parts have been saved, most in tiny remnants. From a wildlife biologist's perspective, the buffalo was a keystone species—a pivotal figure in an abundant and extensive, but relatively simple, community with direct or indirect links to most if not all the members. I've tried to sketch a picture, using the bison as the focal point, of the scene that filled that vast space. Today we can see it only in our minds' eye, and that's hard. I hope this book makes the task a little easier.

I've given a lot of thought to whether I should call my protagonist *bison* or *buffalo*. My scientist side is drawn to *bison*. It is scientifically correct and precise. One kind of buffalo is native to Asia and another is native to Africa; none is native to North America. Yet the side of me that grew up American is drawn to *buffalo*—the name by which most Americans have long known it. I decided to use both names: *bison*, or to be exact *Bison*

bison, places the animal precisely among the world's mammals, while *buffalo* honors its long, intense, and dramatic relationship with the peoples of North America.

One of the lessons thirty-five years of teaching has taught me is that people usually would rather hear a story than listen to a lecture. The buffalo and each member of the community that buffalo are part of have a story. They're great stories, and it's a pleasure and a privilege to be able to tell them.

Some readers will want to pursue some of these stories further. To help them I have placed notes at the back of the book. No superscript appears in the text, but each note (labeled by page number) references a fact or idea, then briefly identifies a source for more information. Full citations are given in the bibliography that follows the notes. This approach also allows me to give credit to at least a few of the many people whose work has made it possible for me to tell these stories.

My research was supported financially by the National Geographic Society, the National Institute of Mental Health, and the University of California Experiment Station. Many people gave generously of their time and talent to help me learn about bison and write this book about them. My research on the National Bison Range was facilitated by Joe Mazzoni and Marvin Kasche and their staffs, while my research on Catalina Island was facilitated by Doug Propst and his staff.

The rewriting of my early drafts was made both much easier and much better by the deep, detailed, and tactfully delivered critiques of my fellow members of the UC Davis Nature Writer's Group—Margaret Eldred, Eric Schroeder, and Kelly Stewart. My colleagues Ben and Lynette Hart, Christine Maher, Tim Caro, Donald Owings, Dirk van Vuren, and Louis Botsford read or discussed specific chapters. Jim Shaw and an anonymous reviewer read the entire manuscript. All provided useful information, insight, and perspective. I took all the photographs except figure 34, which portrays my grandfather and a white buffalo calf; that photograph was taken by my mother, Joyce Lott. My editors, Doris Kretschmer and Dore Brown, and their colleagues at UC Press, Karla Nielsen and Nola Burger, encouraged me, guided me, and helped me through the process of turning the manuscript into a book—making it a better book at the same time. Alice Falk's copyediting of the manuscript was superb.

My wife, Laura, put up with the forgotten household duties, the distractible spouse, and the slipping-sliding piles of books and sheets of paper here and there that my writing always entails.

To all those named above, and to the many more I cannot acknowledge, my heartfelt thanks.

PART ONE

Relationships, Relationships

Bison are gregarious mammals. Each gregarious mammal has a lot of relationships. This section is about bison relationships—what they are, how they're expressed, what they do for bison, and how the animals manage them.

When I first approached bison as a behavioral ecologist, I was in no position to appreciate any subtlety in their behavior. The scientific descriptions of their communication were still very sketchy. Eventually I would fill in some of those gaps. In the meantime the gaps were often frustrating. Yet at the same time it was refreshing, and in some ways more exciting, to be start-ing with a nearly blank slate. It allowed me—forced me, in fact—to experience the behavior of bison directly, not filtered by any expectations beyond a very general understanding of the social and ecological demands and opportunities of being a big plant-eating mammal in temperate North America. Perhaps that's why some of my most vivid memories are of scenes I saw those first summers and springs, and of the thoughts that ran through my mind as I took them in.

At that time I experienced bison social interactions primarily as a grand but largely incomprehensible spectacle, like an opera sung in Italian. I relished the stage, the performers' bigger-than-life presence, the power and purity of their voices. Then, during my first field seasons, I got a rough grasp of bison communication. That allowed me to view scenes at another level—as decoded communicatory behavior creating a rich layer of meaning on top of what I observed. It became like watching an Italian opera with translated lyrics projected as supertitles above the stage. You understand the plot much better, though the distraction of reading the text takes a bit of the edge off your appreciation of the stage and voices.

I've tried to convey some of the drama of bison social behavior in a few scenes without projecting text above them. But the behavioral ecologist's job, and my passion, is to go beyond rhapsodic appreciation by translating the lyrics so that the story line can emerge. Bison social behavior is too marvelous a tale to go untold.

Social life comes down to managing—one could almost say juggling—relationships. For bison the most complex relationships play out during the intense though brief breeding season. Attraction, rejection, acceptance, competition, and cooperation within and between the sexes create vital, compelling, generally short-lived, and shifting relationships. These entangled relationships make one complex story that is told in chapter 1. Chapter 2 describes the simpler, more durable, but equally important relationships cows develop with one another. The final chapter in this section is about the social relationships of calves, especially their relationship with their mother.

Bull to Bull and Cow to Bull

The sky really is bigger in Montana—a colossal inverted bowl of vivid blue. In late July and early August, plumes of dust, rising with earth-warmed air from the brown grass and rolling rangeland, ascend into that bowl. The dust makers, a herd of bison on the National Bison Range, are going about their business—breeding; and I am going about mine—observing, recording, and trying to understand their behavior.

Most of the dust comes from wallows, shallow pits where the bison have torn away the sod with their horns and where the subsoil, dried by the sun and stirred by hooves and horns, turns to a flourlike dust. Some of the plumes start when threatening bulls paw and roll in these wallows; but most occur when fighting bulls plow the soil with their hooves, or when they slam their heads together and the shock explodes dust from their bodies.

Now an old bull bellows. His back arches, his belly lifts, his neck extends, and a sound that seems equal parts lion's roar and thunderclap booms across the grass. An eighteen-inch scar runs up his ribs. His horn tips, shattered in other battles, are blunted and worn. Fifty yards away his opponent, a six-year-old bull in his prime, bellows back, glances at the cow he is tending, then urinates into the dust of a wallow and rolls in it, slamming his 2,000 pounds sideways into the dust. It spurts from beneath him, filling the air around him like the burst of smoke a stage magician vanishes into.

The prime bull emerges from this cloud, headed toward the old bull in a menacing walk. His forefeet stamp with each step, making the hair pantaloons on his legs dance and exploding little puffs of dust from his coat. As each front foot stamps, the bull snorts. His tail stands up like a living question mark. It's an impressive display, and from where I sit, in an ancient jeep, an intimidating one. But the old bull is not intimidated. He too

5

has wallowed and now advances, matching stamp with stamp, snort with snort. As they grow closer their bellows intensify; they seem to signify pure fury.

Most such challenges seem to be elaborate tests of the opponent's determination and end without a fight. Most fights involve a cautious locking of horns or hooking uppercuts or shoving head to head, ended when one animal signals submission and the winner lets him go. But this is not a test of determination and it's a different kind of fight—one of those in which the bulls hurl themselves at each other, elongating their bulky bodies into animated battering rams as they launch themselves for the first blow. Their heads come together with a terrific shock. It ripples through their bodies in a visible wave. I once saw a bull somersaulted backward by such a charge: 2,000 pounds of bull flipped upside down like a lawn chair in a gust of wind.

Both these bulls hold position after the first shock and dig in for a serious fight. They slam their heads together again. Clumps of hair the size of a fist are caught between their short, heavy, curved horns, then sheared off and tossed into the air. The animals circle, each trying to reach the other's flank with a hooking horn. Both pivot around their forefeet with the speed of featherweight boxers, and each parries the other's seeking horns with his own while their powerful necks absorb some of the force of the impact. Their hair absorbs some too. By the time a bull is six years old, a mat of hair several inches thick extends from the top of his head down across his forehead, thinning gradually until it stops just above his muzzle. His eyes peer from shallow wells, his ears flick out from deep recesses, and the space between his horns is completely filled with this luxuriant growth. Beneath this natural shock absorber, a thick layer of tough hide covers a rock-solid skull.

Now the bulls lock horns and push hard, their hooves plowing soil as each tries to drive his opponent back. The old bull is pushed back and a little sideways, dust spurting from beneath his skidding feet. Suddenly a foot catches on a rock and he trips and falls onto his side.

It is rare for a helpless bull to be attacked by the winner, but this time it happens. The younger bull strikes down and forward with his horns, slamming them into the old bull's flank and hooking right and left. The curves of his horns make most of the contact and deliver bruising, possibly rib-breaking, but not fatal blows. Then the tip of one horn plunges through skin and muscles and into his opponent's abdomen. Only one horn penetrates, and it penetrates only once, but the wound will be mortal.

The younger bull ends his attack and returns to his cow. In a few minutes the old bull will rise to walk away. He will graze again, drink again, sleep again. But an infection will send matter oozing down his ribs in a few days, and in a few weeks it will kill him.

Yet as one life starts to ebb, another begins. The victorious bull has mated with the cow he was defending and a calf has begun to form. Its birth is the renewal that has made North America bison country for thousands of generations. Siring as many as he can of that calf's generation is the bottom line for each bull, and it's the imperative that justifies their risking their lives in the battles their bodies are built for.

But while physical prowess is an essential tool in managing a relationship between two males, it can't be the only tool, and in fact it's one of the least-often used. Much more frequently they use communication, which evolved as a means not to transfer information but to do what attacks do—manage another individual's behavior to one's own benefit. In some relationships honesty is the best policy, and communicating animals transfer real information. A mother hen calling "food here" to her chicks is managing their behavior to their benefit as well as hers by telling the truth. They get food, and she gets her genes represented in the next generation. On the other hand, young men in a hormone-induced haze who exclaim, "Of course I love you!" while fumbling with a woman's bra are often lying through their teeth. But both the hen and the men are using signals in fundamentally the same way—to modify another's behavior for their own benefit.

It starts with anticipation. A bull's defenses work only if he knows when and where to deploy them—he must anticipate attacks. So the bull must be a seeker—actively scanning or even probing his environment for clues about whether or not he is likely to be attacked. Call it actively anticipating.

A territorial animal can predict attack pretty successfully by knowing territorial boundaries. The territory owner usually challenges all competitors within a given space and keeps up the pressure with threats and attacks until they leave. But bull bison aren't territorial. They are roamers, drifting singly or in small, temporary groups. Because they cannot use their location in space to predict whether or not another animal will attack them, they read the animals around them, detecting and responding to behavior that consistently precedes an attack. Reading it accurately is a second tool for managing relationships.

It's clearly to the advantage of an animal about to be attacked to become canny in judging his enemy's behavior. Generally the task is made easier by the opponent, who, instead of disguising the coming attack, often amplifies preattack behaviors, draws attention to them, and in every way makes it easy to see what he is about to do. A bull doesn't just walk toward his opponent: he stamps with each step, setting his foreleg pantaloons dancing, and grunting with each stamp. If forewarned is forearmed, why not attack first and give indication later? The reason, of course, is that it may not be necessary to attack at all. Forewarned may be foredefeated—at least often enough to make the warning worthwhile. Call these forewarnings *threats*. A fight avoided is also risk and energy expenditure avoided. Fighting is an occasionally necessary grand spectacle, but the real biological drama lies in the complex, drawn-out, and frequently subtle ways in which most conflicts are settled by communication.

Bulls do most of their communicating during the breeding season— the only time during the year that mature bulls and cows are together for any length of time. The bulls have been alone or in small, temporary groups. Now they join the cows, which have been living in larger groups with the calves and young bulls. The bulls seek out cows about to breed and stay with them (they "tend" them), keeping other bulls away by threatening and fighting.

But threatening and fighting are also common between bulls that are not defending cows. Since receptive cows are the only scarce resource in the bull's economy, this seems surprising at first: one wonders what the nontending bulls are fighting about. But a rival dominated now will probably give way later without a contest, saving a tending bull time and energy when he has none to spare. Not that the bull works it all out in this fashion. He simply has a powerful urge to dominate other bulls, and following this impulse works to his advantage. The drive to dominate is so powerful that it occasionally interferes with his real business and its ultimate function—bulls will sometimes leave a receptive cow to threaten a distant bull.

Virtually all of us air-breathing vertebrates have found ways to turn our exhalations into communications, and most interactions between bulls start with a sound. On a still day a bull's bellow carries for miles. It's a sort of roaring rumble, and if you can't see the bull or don't recognize the sound you may guess it's a thunderstorm. If the competition presses, the bellowing becomes louder, and a quality that is hard to define but somehow easy to recognize—a quality of fury—begins to grow in it. Often one or both bulls will interrupt their bellowing to paw the ground or wallow.

Threatening bulls usually do something we don't understand at all: they urinate into a wallow, then roll in the dampened soil. Do they get enough urine on themselves to send a signal? If so, what could they be signaling? Could they be exposing the opposition to an index of their testosterone level as salient to a bull bison's keen nose as their bulk presented broadside is to his eyes? Or is there another chemical billboard being displayed? As a bull uses up his physical resources during the rut, he eventually begins to metabolize his muscle. The metabolites in his urine will report that chemistry to a sophisticated nose, and the same nose will know when the bull is still burning fat and his muscle is intact. This would be an honest signal of physical condition. The cost of signaling weakness when you are weak would be compensated for by the benefit of signaling strength when you are strong. An intriguing story, but so far 100 percent speculation. Maybe someday we'll know.

If the challenge does not end at the wallowing or bellowing stage, the bulls draw closer to each other and begin to posture. There seem to be two distinct postures. In the "head-on threat," which is simply the posture and movement that precedes a charge, the bull moves toward his opponent with his head held slightly to one side. The more slowly the challengers are moving, the farther to the side their heads are held. When they approach nearly straight on, either one bull submits by turning away or they bang heads. But when they approach slowly with their heads well to one side, they often stop close to, but not quite touching, each other and "nod-threat."

Nod-threatening bulls stand close enough to reach one another; their bodies may form a single straight line or an angle of up to ninety degrees, but in either case they turn their heads aside. From this position they can attack suddenly by hooking a horn into the opponent's head. The hook always starts when the head is close to the ground, the muzzle tucked back. But in the threat itself, the head-low, muzzle-back position is only a brief interruption of a head-high stance: the bulls' heads drop in a matched movement, then swing back up again, still to one side. A hooking attack may start at the bottom of any one of the down swings, but the opponent never seems to be caught off guard. After a series of such nods one animal may suddenly submit, ending the clash.

Nod-threatening takes place most often between bulls that are not tending cows, as does the "broadside threat." A bull in this posture keeps himself broadside to his opponent with his head held a little higher than normal. Usually his back is arched and he is bellowing. If he moves, he does so slowly, in short, stiff steps that keep him broadside to his opponent. Often two bulls will threaten by standing parallel to each other just a few feet apart. Only rarely does this threat lead to a fight. The encounter may be long as threats go, lasting up to a minute or more, but one of the animals almost always submits.

The broadside threat and the nod threat emphasize the degree to which the bulls forewarn their opponents. This forewarning is so elaborate that it has become a force in its own right. It goes beyond permitting the prediction of attack: by substituting for attack, it often overpowers the opponent.

That function may account for some puzzling aspects of these postures. Why, for example, do the bulls threaten by turning broadside? When turned this way, a bull seems very vulnerable to attack, particularly if the bull he is threatening is facing him. Since all his protection is concentrated at his head, the usual point of attack, a bison bull is easily wounded by a horn thrust from the side. (This danger, by the way, is more apparent than real. In watching thousands of such threats I have seen only one attack on a bull turned broadside.) Perhaps the function of the broadside threat is to display the full size and power of the bull, as well as to forewarn the opponent. If the threatening bull makes a big enough impression, he may save himself a fight.

The one recurrent note in all these descriptions of fighting and threatening is that they go on until one animal submits. Submission signals have two functions: they enable a bull to withdraw from an encounter without getting into a fight, and they enable a losing bull to end a fight without retreating a long way. There are two questions to ask about bison communicating submission: How do they signal it? Why do the winners accept it?

All bison submission signals are variations on a theme: the submitting bull turns away. Sometimes it's a 180-degree turn followed by a galloping retreat. At other times it's an abbreviated swing of the head and neck to one side. When it involves a 90-degree turn, the submitting animal ends up in the same general position as one who is threatening broadside. But it's easy to tell the difference. In submission the bull's head is usually low, muzzle extended as if to graze—and sometimes he does graze—and the bull is silent. Whatever form the submission signal takes, it almost always stops the threats or attack immediately.

One day two bulls dramatically demonstrated the power of this signal. They were fighting in a swale below me. The low ground was moist, so the grass was green even in early August. The spurting dust raised by most fights was missing, and the rich contrast between the warm brown of the bulls' coats and the green grass gave the scene a certain tranquillity. But the bulls were fighting in earnest. They slammed their heads together, stepped back a few feet, then drove their foreheads together again so hard that the shock of the impact seemed to pass visibly through their bodies. After three or four such blows, just as they had drawn back and were poised to plunge together again, one of the bulls simply stood in his tracks and swung his head ninety degrees to the right. The winner had already started his forward lunge. His front feet plowed sudden dark furrows in the green grass as he skidded to a stop. His horns could not have been more than eighteen inches from the loser's neck. The two animals stood immobile for a few seconds; then both walked quietly away.

Fights usually stop just that abruptly with the loser just that vulnerable. One more step, one more lunge by the winner, and the loser would be out. But that step is almost never taken.

Why does an act of submission change the winner's behavior so profoundly? One is tempted to explain it by analogy to human institutions, to say that bison operate by a set of "rules." Thus the loser is kept from harm by the winner's acquiescence to rules, just as a football player who has been knocked off his feet is protected from further assault by the rules of football. But this kind of analogy between social behavior and social convention obscures rather than clarifies. The rules of football are a social invention based on enlightened self-interest and reciprocity. Players agree to be restrained from some destructive acts, provided that the opposition is similarly restrained. Kicking an opponent in the head when he is down is so dangerous that everyone agrees that it should draw a penalty that gives the other team some advantage in play. This penalty imposed by a specialized group of rule enforcers is the mechanism through which rules control the football players' behavior.

There are, however, no reciprocal agreements and no rule enforcers among bison. Each bull's behavior meets his own needs and no other bull's. The only penalties for any action are those assessed by the action itself. At first this seems wrong. How would a winning bull penalize himself by polishing off a loser? In fact, he would be deprived of two precious commodities: time and energy. The breeding season, when most conflicts take place, is limited, and time spent fighting, even a mop-up operation, is time lost from breeding.

Fights to the finish would take even more energy, and that's in shorter supply than you might imagine. When you see bulls in the middle of summer, in the midst of tall grass and warm sunshine, their good health and nutrition seem assured. But bison are northern animals, one of the most northern of the cattle family. They have adapted to a climate where food is scarce through long winter months. Bulls die during the winter if fall catches them without enough energy stored as body fat. As it is, the breeding season takes a lot of energy. Mature males lose an average of 200 pounds between June and October. If every fight were long and rough and ended in a cross-country chase, bison bulls, winners and losers alike, might well die before spring renewed the plains.

Pursuing and destroying the loser would eliminate the need ever to face him again, of course, but that would accomplish little. Bison breed in large groups, with the males moving constantly from one group to another. There are always many more challengers where the last came from. In the final analysis, the winner does not spare his defeated rivals, he spares himself.

The prolonged forewarnings, the reluctance to fight, and the generosity to losers are neither the last noble vestiges of chivalry in our time nor nature's way of exhorting humans to live on a higher ethical plane. Rather, they are carefully balanced behavioral adjustments to the social and ecological circumstances in which the competition between bulls evolved.

OLDER AND BOLDER

Picture this: two men, strangers, both sober, are quarreling about which will buy a drink for a young woman who sits on a stool watching. One is old enough to be subjecting strangers to baby pictures of his first grandchild; the other is just old enough to drink legally. One of the men is pushy—escalating from words toward violence; the other is cautious and restrained. Now, if you can, picture this—it's the old guy who is escalating—shoulders hunched, on the balls of his feet, fists clenched, movements abrupt, voice loud and angry, while the young guy is back on his heels, hands held up and open before him, trying to avoid a fight.

There's a lot wrong with this picture. It's a rare young man who gets bolder as he gets older. But the ecologists Chris Maher and John Byers have shown this to be the rule in bison. The older they get, the more likely they are to take the risks of combat in order to tend a breedable cow—not because older mature bulls are more likely to win than younger mature bulls, but because they have less to lose. Thereby hangs an intriguing budgeting issue, which might be titled "How to spend your life the way that buys the most descendants." It's an investment program in which your life is your capital and the return is your offspring's offspring. For male bison, producing offspring usually involves conflict. Each such conflict puts their lives at risk. A bull bison's optimal investment strategy weighs possible losses against possible gains and decides how much risk is prudent. He's balancing making a killing against getting killed. Now, a male is not going to live forever, so an optimal strategy must be age sensitive. The younger he is, the more time and future opportunities he stands to lose if he dies. So he should invest cautiously when he's young and has more to lose, but more and more boldly as he gets older and has less to lose. Bison behavior tracks this straightforward logic—the old bulls are the bold bulls.

But a man finds such budgeting more complicated. For the bull, getting his sperm on the way to the egg is the end of his investment. For a man, it's the beginning. Until the offspring reproduce, the father's investment has not paid off. Human babies don't stand on their own two feet within ten minutes of being born, don't run in two hours, aren't weaned at six months. Children must be fed and protected for years, and fathers can help, thus investing more in each child. As the investment in children grows, so does their value. Once he has children, staying alive to support their development and eventual reproduction and maybe even that of their children's children is a good alternative to risking his life to produce another child he may not live long enough to rear.

Small wonder grandfathers would rather dote than fight, and are more likely to risk their lives to preserve their grandchildren than to win wooing rights to a woman, no matter how fecund. This attention to one's young is called *nepotism*. Bull bison never engage in it, but human males, whose genetic survival such lofty detachment would threaten, engage in it all the time—occasionally despite a whole body of civil service law invented to thwart their urge to do so.

Not only is life a series of trade-offs, but life itself is one of the things that gets traded off. The trade-offs that natural selection selects usually bring all the costs and benefits to bear and therefore favor very different strategies in different kinds of animals. Different strokes work better for different folks.

COW-BULL RELATIONSHIPS

Bison bulls, and many bison watchers, are preoccupied with bulls' relationships to other bulls and with the tools—spectacular and subtle—they use to manage their relationships. But such tools are just the means to a more important end, an essential and showy precursor to what matters most: making calves.

Many living things reproduce with neither preliminaries nor fanfare. They just split in two, grow a bud that becomes a new entity, send up a shoot from a root whenever and wherever. They don't need any help, or even cooperation, from another individual. But apart from a few parthenogenic species of fish and lizards, we vertebrates are stuck with sex. That means, at a minimum, two members of the species being at the same place at the same time so eggs can attract sperm and get fertilized in a fertile egg–friendly environment. For some fish the water around the coral reef where they live is plenty good enough. One to hundreds of females get close to one to hundreds of males, each sex releases his or her gametes at the same time, and eggs and sperm find each other and fuse.

But for mammals the equivalent of warm seawater is in a female's uterus. The eggs stay there and the sperm have to journey to join them. This journey requires some seriously specialized equipment and some seriously intimate contact. Where this contact occurs isn't very important, but when and with whom matter a lot. Cows, unlike coral reef fish, release one egg at a time. To get that egg fertilized right they need to attract a suitable bull and become willing to mate with him. Neither of these things is everyday activity for a cow, so she requires substantial changes— changes that make her more attractive to bulls and, eventually, changes that make her willing to breed with one of them.

Back to bulls for a moment. The way to a bull's enthusiastic attendance is through his vomernasal organ, a sensory organ found in many mammals; it has an opening in the roof of the mouth. Cows' urine is full of facts about how near ovulation is. The bull's vomernasal organ seems specialized for an analysis of female urine chemistry that provides information on when a female will be ready to breed. But while bulls will joust seriously to get some female urine on their vomernasal organ, I've never noticed that females go out of their way to present it. However, I have many times seen bulls during the rut bring a resting cow to her feet by prodding her belly gently but firmly with a horn. The cows arise, with what I take to be resignation, and often urinate in a minute or two. The bull thrusts his muzzle into the stream of urine, then elevates his head, upper lip curled, tongue fluttering inside his mouth, his whole demeanor suggesting a gourmet's appreciation of a fine wine. If he goes from lip-curling to tending, the chances are good that the cow will breed sometime that day.

There are two phases to a cow's getting-pregnant physiology: she must ovulate and she must become willing to breed—enter estrus. Each phase is the end point of a complex sequence of hormonal events, and chemical traces of these events appear in the cow's urine. Since each sequence takes several days, chemical signposts show up in the cow's urine days in advance. We call the several hours before the cow breeds *pre-estrus*, and bulls that test the urine of pre-estrous cows are likely to try to spend the next few hours with her. That's not easy to do. Pre-estrous cows become restless—breaking away from a tending bull and running through the herd. A running cow attracts bulls, and a string of them are soon following her just as a tail follows its comet. When she stops they gather and quickly sort out who among the present company gets to stand by his cow. The cow's best shot at having many grandchildren is to have sons that can claim a cow just as this bull claimed her. If we assume "like father, like son," he is the best candidate in the immediate circle—but the cow may well make him prove it again with another run through the herd.

I always interpreted cows' runs as simply what they did at a particular stage of estrus. The zoologist Jerry Wolff thought otherwise, and he gathered data showing that the lower the tending bull's rank, the more likely it was that the tended cow would run. In addition, cows that ran usually ended up with a bull that ranked higher than the one they ran from.

It's just possible the urine left by wallowing bulls plays a part in the cows' physiological progress toward ovulation. All bison wallow and frequently sniff at a wallow before rolling in it. The smell of male urine could stimulate ovulation and a breeding frame of mind, as it may do in moose. Of course a cow can ovulate on her own, but not all ovulations are equal. If you're a calf born on an open grassland roamed by predators, the best cover you have is other newborn calves, who may fill the predators' bellies before they get to you. The biologist Richard Estes has shown that wildebeest calves born at the height of the calving season, when the African savanna's predators are relatively satiated, are the least likely to be eaten. So using a clue, any clue, that other cows are relying on to time ovulation can improve a particular cow's chances of having grandchildren.

We don't know if bull urine helps synchronize cow ovulation but we do know that cow urine launches bull-cow relationships. A pre-estrous cow's chemistry attracts one-year-old bulls as powerfully as the ten-year-olds, and they all vie to tend a pre-estrous cow.

Tending pairs are unveiled as the movement of a grazing herd leaves them behind. Study a pair and you will see the cow grazing a bit, looking fretfully toward the increasingly distant herd. A big bull stands beside her, moving to block her when she sets off after the departing herd. He moves like a basketball player staying between a dribbling point guard and the goal. Sometimes she allows him to hold her in place, but "allows" is the operative word. His moves to block her are quick and graceful, but she is still quicker and more graceful. If she stays it's because she has chosen—or at least settled for—him, not because he has chosen her.

She may head straight for another tending bull. When Jerry Wolff compared the ranks of the bulls left to those approached, he found the cows were usually approaching a higher ranking bull. That's one of the forms of choice a cow has, and it makes sense for her to be as choosy as she can manage to be. She will have just one descendant in the next year, and only half its genes will be hers; she should be as fussy as possible about where the others come from. But the bulls usually severely limit the cows' choice. On any given day there are fewer ready-to-breed cows than eager-to-breed bulls in a herd with a natural sex ratio—that is, nearly as many bulls as cows. The bulls use their bull-bull relationships and social tools to allocate this scarce resource. But the cows are not simply passive. Cows seem to be more receptive to older bulls, and that makes sense; winters, battles, disease, and predators have tested them.

We don't yet know what cues besides age the cow may use in making her choice. Could all that bellowing make a difference as to which bull is standing beside her when she stops running? Could it work like some birdsongs or frog croaks—a clue to the female about who might be a better mate? The bulls don't seem to be sending a signal—they appear only to bellow to other bulls. They seldom bellow unless they already have a cow, are trying to displace a bull that has one, or are in the midst of a dominance contest.

Even so, the cows can't help but hear them, and I have seen them react. One day I watched a closely tended but resistant cow standing quietly beside the tending bull. Her tail was clamped firmly over her vulva, she chewed her cud, her ears lay passively back, and she jumped away from the tending male's attempts to mount. He was bellowing and glowering at a half dozen bulls standing in a semicircle around the tending pair. It was an all-around stalemate. Then another bull broke it. He walked in from directly behind the tending pair, and when he had closed to forty feet he bellowed. The effect was electric. The tending bull left without a backward glance, hurrying to join the semicircle, and the cow, without a backward glance, lifted her head and tail and flicked her ears forward.

The new bull continued past the cow, stalking stiffly around the semicircle while the bulls forming it ducked their heads down and away as he passed each of them. As he presented his right side to the semicircle of bulls the cow pressed against his left side, half mounting him every few steps. When he stopped after passing the last deferent bull the cow stepped directly in front of him, lifted her tail and braced herself. He mounted immediately and four or five seconds later her breeding season was successfully concluded.

However important bellows may be to a resistant cow, her priorities and behavior change as ovulation approaches. The time of choosing is a period of conflict between the cow and the many courting bulls. The more competitors, the better for her but the worse for him. His earlier behavior minimized the number, hers maximized it. But now their interests coincide, and they must cooperate and collaborate.

Often the cow redirects the relationship. Her physiology is changing fast, altering her behavior along with it. Now, instead of breaking into a gallop every time the bull is distracted by a challenger, she follows him, and when he has disposed of the distraction she stands close, perhaps even positions herself in front of him. If he continues to glower round at the competition instead of mounting her, she may announce her readiness to breed by licking him or by sticking a horn in his ribs and prying upward—extracting from the bull a grunt, a tuft of hair, and more attention. She may even mount him, and when he begins to mount her she no longer squirts forward like a stepped-on bar of wet soap, but plants her feet and moves her tail to one side. Even so he may half mount, then drop off several times before he catches on and copulates.

Bison sex does not involve a lingering mingling of mucous membranes. He clamps his forelegs around her ribs and penetrates with a lunge. The bull almost always ejaculates within five seconds of intromission (I timed it from movies of the event). His last pelvic thrust is driven home by a contraction of his abdominal muscles so strong that it jerks his hind feet forward and completely clear of the ground, making the 1,100-pound cow's hindquarters suddenly support an extra ton of buffalo. Brief though the encounter is, it's usually enough for the cow. She staggers under his weight, not infrequently limps for a while afterward, and four times out of five rejects further attempts by this or any other bull to mount her again for the coming twelve months.

What a difference five seconds can make! In five seconds the cow is transformed from eager to unavailable. In just five seconds she has gotten everything of value the bull has to offer for the coming year. Only one cow in five—almost always a cow that bred near the end of the breeding season—will stand for another mounting during one breeding season. In fact one is enough. In a well-nourished herd, 85 to 90 percent of the mature cows will bear a calf in the spring.

Though she moves a bit gingerly—back arched, tail extended, sometimes limping—she is soon grazing again. Her breeding program is completed, but not the bull's. His season is in stride, and its success depends on two things—being the only bull to fertilize each of the cows he couples with and coupling with as many cows as possible. Just now a dilemma has him on *its* horns. This cow could be one of the 20 percent that accept another mount. If she is and he has left her to search elsewhere for a fertile field to plow, her second coupling will be with another bull whose sperm will compete with his for the one egg headed for the cow's uterus. If he stays with her, he will be the bull at the second standing and all the sperm will be his. He can ensure this by staying with her (or, as behavioral ecologists say, "sequestering" her). But the longer he stays, the more likely the other cows' dance cards will already be full when he comes courting. Breeding season moves at a spanking pace. I've seen half the cows breed in the peak four days. To be here or to be there, that is the question. The answer is, play the odds. The odds are that a cow that breeds a second time will do so within forty-five minutes of the first encounter. Most bulls play the odds, sequestering the cow for about forty-five minutes, then moving on, looking for a new relationship.

It's enormously taxing. While breeding, bulls lose 10 to 15 percent of their body weight—mostly fat they will dearly miss in the coming winter. But the potential rewards are also enormous. The winning bulls win big. One year I saw one bull breed five cows while others bred none—one-third of the bulls sired two-thirds of the coming spring's calves. Over three years Jerry Wolff saw one bull breed sixteen cows, while another never bred. The biologists Joel Berger and Carol Cunningham followed a herd for four years and saw one bull breed twenty-eight times while others never bred.

Breeding season ends with neither a bang nor a whimper, just a fairly rapid decline in the number of tending pairs. And more and more of the bulls drift away to concentrate on providing their complicated digestive system with the fodder it will convert to fat to carry them through the winter.

Their interactions change from constant confrontation to nearly invariable tolerance or passive avoidance. They're not looking for any trouble, and, in the two months following the rut, they come to look a lot less like trouble. The magnificent, menacing mass of hair on their forehead and between their horns, the flowing beard, and the dancing pantaloons that gave advancing bulls such presence are gone. Much of the hair between their horns was barbered away—caught between rubbing horns and sheared off during fights. But the rest of that hair, the beard and the pantaloons, simply falls out after the rut ends and before winter starts. Only mature bulls molt this way, not cows and not even young bulls.

Perhaps the hair loss is triggered by the stress of the rut—and perhaps instead, or in addition, it de-escalates the tension between the bulls: each benefits by looking less big-male threatening and thus less likely to provoke other mature males to challenge him, a management tool worth having when your goal is to graze as much and exercise as little as possible. Then, as winter wanes, the beard, the pantaloons, and the hair between the horns regrows and the bull again wears his special combination of parade ground display and battlefield combat dress. He will again be able to deliver or to survive a full gallop charge squarely on his forehead, but still won't if there is any other way to manage his fellow bulls.

BULLS IN SPRING

The midsummer breeding season is full of sound, fury, dust, and danger. But it would be suicidal to try to keep that up all year round. Bull-bull relationships in the other seasons are comparatively understated. In spring the bulls are hurrying to get ready for summer and winter, but they hurry, in good part, by taking it easy.

Two old bulls are lying down, resting and ruminating, about thirty feet apart. The nearest other bison are at least a mile away. One bull rises, stretches, and stands, and in a few minutes the other does too. The first to rise approaches the second. At a distance of fifteen feet, a little tension develops. Their bodies stiffen slightly, and their tails, which had been discouraging flies by flicking from side to side, become still. The bulls look carefully at each other, then slowly relax. They lick their own noses, sending their tongues into first one broad nostril, then the other. They amble off companionably for a hundred yards, grazing as they go, then lie down again.

In the summer, in the breeding season, they will be competitors. They may be locked in combat, even mortal combat. But this is no time to quarrel. This is a time to put on fat against the demands of the breeding season in summer and of the cold winter that will soon follow. Not all animals, not even all hoofed animals, must obey this imperative to get fat once a year. Neither the African buffalo nor the Asian buffalo ever get very fat. The seventy species of African antelope don't either, but they don't have to contend with winter. Most animals that do, like the bison, live according to a "fat economy."

This strategy simply means that the animal builds up (saves) a store of fat when food is abundant, then lives off it when food is scarce. Both black and grizzly bears are archetypal fat economy animals; they pile up fat during a few frantic months of gluttony, then hibernate during the winter, resting cool and even comatose while life is sustained by a store of fat melting slowly like a candle's wax. Storing and using fat is somewhat inefficient. Converting digested food to fat uses some of the food's energy, and converting it back to usable energy requires still more. But what the process lacks in efficiency it makes up in reliability. It makes it possible to survive lean times like the hard winters that always lie ahead of bison.

Spring is the time of year for fat economy animals to eat. Not only are their fat stores depleted, but the chance of replacing them is best then. Fat is stored energy, and for most living things energy ultimately comes from the sun. Plants convert the sun's energy into a form plant eaters can use. Though plants are stationary, they too are in a race, a race to have the most surviving seeds or stolons. And so they grow aggressively, sometimes with their greater size depriving their neighbors of sun or water. To the herbivores these weapons are the means to get fat. In the spring plants are not only abundant but also especially nutritious. Growing grass and the flowers, seeds, and other reproductive organs of plants are rich in protein, a critical nutrient that is generally scarce in grass. So now, as the energy from the sun increases day by day, primary production surges and the bison can get fat.

And so today the two old bulls move on, grazing, resting, ruminating, and grazing again. Tomorrow both may be all alone and miles apart, or one or both may have joined three or four others; but wherever they are and whoever their companions, they will be grazing, resting, and ruminating: making fat while the sun shines. Their slow pace and laid-back social behavior are their way of hurrying, the fastest way to their goal—fat stores.

CHAPTER 2   Cow to Cow

DOMINANCE RELATIONS

Buffalo cows often seem as contented as the legendary Elsie. A small group grazes green grass, tails swinging languidly; heads nod gently each time a mouthful of grass is clamped between lower teeth and upper hard palate, then torn free with a little jerk. The group moves together, heads all pointing the same way, one cow or another pausing to lick a ticklish rib and look around, a few feet from the cow to the left, a foot or two further from or closer to the one to the right. Serenity personified—with big eyes and small horns. Well, why not. Bulls have to compete to mate, cows don't. Cows need grass to eat, water to drink, companions as a buffer against predators. But food can be in short supply. Whenever a resource is, as ecologists say, "limiting," it's worth contesting, and a social system that allocates that resource will emerge via natural selection. It wouldn't pay for cows to defend territories, so we should look hard for a dominance type of social system.

After the roaring, thundering battles of the bulls, the conflicts of cows seem mellow and understated. For one thing, aggression at the level of head banging is rare. The zoologist Alan Rutberg kept count of cow disputes on the National Bison Range. Let watching two cows for one hour equal two cow-hours. In some 26,000 cow-hours, he saw three fights (horn clashes) and one broadside threat. In some 6,000 bull-hours during one breeding season there, I saw 123 fights. Even when they're getting serious, cows' clashes can seem more comic than cosmic. I've seen cows urinate thousands of times and wallow thousands of times, but only once have I seen a cow put urinating and wallowing together as a threatening bull would do. It wasn't breeding season. A cow group was grazing quietly when a lone cow walked up and started to join it. One of the oldest cows walked out to meet her, pawed in a wallow, then urinated and wallowed. Her sequence was just like a bull's, but her geometry directed the urine to the grass behind the wallow, while the wallow and the wallower remained perfectly dry. The old cow's belligerence kept the would-be joiner at bay for perhaps fifteen minutes; then she went back to grazing while the newcomer quietly joined in.

Although the old cow's urination-wallow was a touch comical, dominance relationships between the cows are no laughing matter. While there's very little violence, there's lots more subtle action. Alan counted aggressive interactions and saw two per cow-hour, ranging from a subordinate withdrawing to a dominant swinging her horns or lunging. The cows are under social pressure, expressed in the physical and social distance between neighbors. As is so often true of participants in relationships, they want to be close, but not too close. The closer a cow's neighbors, the less the danger from wolves but the more the competition—for food, wallows, and water. Being close takes a lot of fine-tuning. Every step brings you closer or takes you farther away from several others, all of whom are also fine-tuning the distance they are from you. Physically, a buffalo cow just plods across ground and through air. But socially, it's as though the air between each two cows were contained in transparent bags that compress and expand as each animal moves closer to or farther from a neighbor.

The pressure inside the bags depends on the relationship between a cow and each of her neighbors; and since the relationship is not symmetrical, the pressure of the "same" bag is different for each of the cows. The dominant member of a pair may feel a little pressure when stepping toward her neighbor, while the subordinate may feel intense pressure when stepping toward the dominant, and not move as far. In addition, the cows can vary that stable, baseline pressure. A deferential head duck by the subordinate decreases the pressure for both; a threatening head swing by the dominant raises the pressure for the subordinate. The cows aren't seeking an absence of pressure. If they can't feel any there they find somebody. And up to an optimal point, the more pressures they can feel the better.

Being a dominant member of a group has high potential payoff. Alan Rutberg broke the process of feeding into two parts: searching and harvesting. In Yellowstone Park, subordinates searched more and harvested less than dominants. Feast and famine were regular visitors to the Great Plains, with wet weather in some years and drought in others. In drought years every mouthful became precious to a cow eating to store enough fat to trigger ovulation and to carry a growing fetus through a hard winter. Dominance would have a big payoff in those years. And dominance would pay off any time that the snow was deep. Foraging bison sweep the snow from the grass by swinging their heads from side to side and using their muzzles as plows. They clear little craters in the snow and eat the grass at the bottom. At least, they eat that grass unless and until a dominant makes them move on and takes over that patch. A dominant will eat everything it clears and some that it doesn't clear. A subordinate will eat only part of what it clears. That difference can be really big at crucial times. All other things being equal, the dominants will get fatter and the subordinates will get leaner. So the cows are interacting constantly, dominants and subordinates in a careful dance of distance; and being the dominant is worthwhile.

A socioecologist can't resist asking what makes one of a pair dominant and the other subordinate. The answer turns out to depend on where you ask the question. Alan Rutberg asked it at the National Bison Range. These particular cows were, as my dad would have said, in grass up to their bellies. Every cow was well-nourished, and every year about 90 percent of the mature cows produced a calf.

Alan worked out which cows were dominant in relation to which. He found that two-year-olds dominated one-year-olds, and three-year-olds dominated two-year-olds; through age eight, all older cows dominated all younger ones. When they were young, of course, the older cows were also bigger than the younger cows. But by the time they were three years old that was no longer so, for they'd reached their full size. Yet three-year-olds, even when bigger, didn't dominate six-year-olds. It appeared that there were no upstarts because there was nothing to start them up. In the midst of abundant food, they had nothing to gain and they just laid low. Among adults there was no relationship between weight and dominance. What mattered was ecological resources, for which animals compete when necessary—and dominance status is the coin of competition. It does seem odd that Alan saw two aggressive encounters per cow per hour if these cows weren't competing for scarce resources. Yet it's typical of a dominance system that the dominant individuals launch a steady stream of preemptive bullyings—sort of like those "Don't even *think* about it" admonitions.

But buffalo cow society was very different on Catalina Island. Once severely overgrazed, Catalina, despite decades of restoration, offered slim pickings to the 400 head of bison grazing there; and that raised the stakes for the cows. At the same time that about 90 percent of mature cows on the National Bison Range gave birth each year, on Catalina only about 30 percent annually were giving birth. We captured seven cows in a corral, determined their weight and age, and then hung radio transmitters around their necks and watched them. In a corral the cows had a strict dominance hierarchy. That hierarchy correlated perfectly with body weight: every cow dominated every lighter cow and was subordinate to every heavier one. Weight was all that mattered—age was irrelevant. Dominant cows ate better than subordinate, going through the oat hay in the feed troughs to get the grain that fell from the shattered heads and leaving the straw for the less dominant who were waiting their turn.

After a few days of watching, we turned the cows back to the Range and followed them for four years. They went their separate ways, and we only rarely saw them together again. The more dominant they were in that corral for those few days, the more calves they had. The heaviest, most dominant cows calved every year; the lightest, most subordinate not at all. That is a huge difference. Natural selection is nothing more than some individuals rearing more offspring than others. If eating better produces more offspring, then any trait that supports better eating will spread. And we know that at least sometimes dominant cows eat better than subordinates.

So it seems that on Catalina, where dominance pays off, cows compete for it. And rather than order of birth determining the outcome, there is a perfect correlation between cows' weight and their dominance status. It looks as if scarce food raised the level of competition by making dominance worth taking more risks and spending more energy; either heavier cows became dominant or dominant cows became heavier.

Bison cows are thus no Johnny one-notes. Their forebears must have experienced both plenty and scarcity in their evolutionary history—times when dominance was worth fighting for and times when it wasn't. Changing circumstances select for changeable behavior, with different strategies for different times and places. The bison pay the costs of striving for dominance when the benefits are high, and don't when the benefits are low.

CHAPTER 3   Cow to Calf

On the National Bison Range calves are born in April and May—spring fever season. The snow has melted and the earth is warming. The new grass growth's vibrant green is eclipsing the brown, dried grasses of winter. The golden yellow of arrowleaf balsam root and the purple of lupine contrast intensely with the new grass. I have watched dozens of calves emerge into this idyllic world. It's a setting to encourage relaxation, even lassitude—I feel it, but the calves don't seem to. Take the one I'm watching just now. Within a minute or two, as soon as his mother has freed him from the membrane that surrounded him in the womb, he begins a frantic-seeming struggle to get to his feet. He gets halfway up several times, and falls forward, backward, and sideways. I think, "Take it easy, little one, rest a minute. There's no rush!" But the brain that has guided calves to adulthood for thousands of generations knows better. The calf hasn't got a minute to spare. Wolves may arrive any moment. A late winter storm could drop six inches of snow tonight, and winter is certain to return in a few months.

Winter and wolves. These ancient forces selected which among calves past would bear or sire another generation. And so they have shaped this calf—bones, brains, and behavior. To survive them the calf must grow: bigger, faster, fatter. So much growing to do, in so little time. The calf responds to these imperatives as to whispers from the unbroken chain of ancestors who obeyed the commands and lived to pass them on: "The wolves will come. Stand and grow stronger. Exercise and grow fleeter." "Winter is coming. Eat and grow longer legs before the snow is deep. Eat and grow a bigger body that will conserve heat in the winter's cold. Eat and store fat for energy when food is short and nights are long and cold."

So the calf lives to grow and grows to live. It must stay alive to be able to grow, it must grow to be able to stay alive. That is its work at this stage of its life. We're tempted to think of the calf as an incompetent adult, especially when we see a male a few weeks old charging about butting heads with animals twenty times its size, or enthusiastically but fruitlessly mounting a female calf. But its work is not to be an imperfect adult (or any other kind of adult); it is to be a perfect calf. The same work any other developing creature has.

Ruminants demonstrate the difference in the work to be done with rare clarity. Their family name comes from the practice of bringing up fist-sized wads (boluses) of partially digested plants from the fore part of the stomach and ruminating: meditatively chewing, then reswallowing, the boluses. Adult bison spend a good part of their day ruminating; it's an essential part of their digestion. But bison calves don't ruminate for the first three months. Not because they are imperfect adults, but because they don't yet get their nutrition from plants. At that stage they are nonruminating ruminants, which is exactly what they need to be. Mother's milk makes this way of life possible, and the relationship between the cow and the calf is what makes mother's milk available.

A buffalo's first breath, its first sight of light, comes as it emerges from a warm womb into the brightness of a spring day or the dimness of a spring night. It's not alone. Mom is there, a bit tired from her labor but ready to begin a foundational relationship—mother and child. The calf is still tangled in the membranes that enclosed it in its mother's womb, and the tattered, bloody remains of the placenta and related membranes dangle from beneath the cow's tail.

Mother is avid for her baby. She licks, pulls away, even eats the membrane that entangles it. When the membrane is gone she licks the drenched coat beneath it. Love at first sight comes only to those who are ready for it, and this mother is ready. As she gave birth, oxytocin, a hormone from her pituitary, surged into her bloodstream and onward to her brain as it does in every mammal, humans included. As her womb emptied, her heart filled and her senses absorbed the sight, sound, taste, and smell of what would be, for the months to come, her one and only. The one living thing she will nurture—and sometimes even challenge hungry wolves to protect.

The calf needs to stay close to its mother, and to nurse. Evolution not only made nursing a necessity but set up a positive feedback loop. Nursing releases oxytocin, so the more the calf nurses the more mother loves it; and the more she loves it the more she allows it to nurse. She even lets it interrupt her travel when it presses its body across her front legs, then dives for her udder when she pauses.

A bison calf's first priority is to get to its feet and walk. By the stopwatch I kept time with as I watched a dozen births, that takes all of seven or eight minutes. An hour-old calf can scamper pretty well. It's a wonderful adaptation to being born in plain sight on a prairie with wolves about, but it makes the calf a challenge to keep in touch with—it's like a ball that never stops bouncing.

The bouncing baby bison doesn't bounce aimlessly. It bounces toward something big and close. Mom is big and close and the ball usually bounces her way. But sometimes it fixes its eye on some other bison that passes by and rushes after it. Then mother chases both down and retrieves her young. A calf a few months old that loses its mother will attach itself to anything large and moving. An orphan calf followed Captain Meriwether Lewis all one afternoon as he walked west beside the Missouri River.

Most hoofed babies are like baby bison—on their feet and ready to move the day they're born. We say they're *precocial*. Some species stay with mother from birth on—we call that the *follower strategy*. In some species mothers hide their babies and leave them, returning only to nurse them. We call that the *hider strategy*.

Pronghorn are a hider species. Pronghorn fawns are odorless, and small enough to disappear by lying down behind grass ten inches tall. With their coloring and curled-up posture they bear a striking resemblance to a large, dry buffalo chip when viewed from above. Mother returns to the area periodically, stops near the young—nearly always twins—and calls. The fawns jump up, rush to mother, and nurse. After several minutes of enthusiastic nursing the young walk off and lie down fifteen or twenty yards from the end of mother's scent trail. This way a coyote can't follow a mother's scent to her fawns.

Follower versus hider is a very useful distinction, but more as a way to identify basic strategies than as a classificatory scheme into which the particular behavior of particular animals must be forced. In the latter case it tends to blur the middle ground. The research biologist Wendy Green has pointed out that yes, bison calves are followers, but unlike some young ungulates who stay at mother's heels, often they're following the herd as much as their mother. Being in the herd has one of the advantages of being with mother—it's safer than being alone. The herd's adults might challenge wolves. Even if they don't, there's a good chance some other calf will be between you and the wolves. That means you don't have to outrun the wolves, you just have to outrun that other calf. A big downside is that you may lose contact with mother if the herd stampedes.

For bison, as for people, keeping in touch can be figurative as well as literal. Many animals that move in herds or flocks have a call that signals with much the same effect of a human crying out, "I'm here; where are you?" The bison version of this "contact call" moved my veterinarian grandfather, when he was superintendent of the National Bison Range, to refer to the bison as "my pigs."

You hear their call occasionally while the animals are grazing, more frequently as a group of cows and calves walk along as they're going somewhere—say to water. You hear a lot of grunting when a herd has just stampeded, separating cows and calves. Mothers and calves grunt to each other across a poststampede herd and track the right-sounding voice they hear to a reunion. The grunts that are so alike to our ears are different enough to theirs to convey identity—like a familiar voice saying hello when answering your phone call. But like someone responding to an on-the-phone hello, the hearer sometimes gets it wrong. I've several times seen a cow and a calf exchange grunts across a herd, make their way through it and come together, noses extended, only to fail the gold standard—the sniff test. Away they go, grunting again and listening again, for *the* grunt.

This too shall pass, but not for several months; and while it lasts it is the most intense relationship of a bison's life. The intensity will subside with weaning, sometime after six months, but the relationship, particularly if the calf is female, can last much longer—can, but doesn't always. To measure the life span of a relationship, you must record first its birth, then its death. It's easy to say when a cow-calf relationship begins, but harder to say when it is over. So we rely on proximity—inferring that the closer the bodies, the closer the relationship. But how close is close? I used a pretty crude measure to follow cow-calf relationships among the 400 head on Catalina Island. Cows and calves there are usually in small groups. My graduate students and I disregarded anything subtle going on within groups. We simply determined how old the calf was when it became so independent of its mother that it was no more likely to be in the same group with mom than it was to be in a group with any other individual.

The bull calves we watched were no more likely to be with their mother's group than with any other group from six months on. The heifer calves were in the same groups for a year, then they too were independent. Yet the daughters that Wendy Green followed at Wind Cave National Park in South Dakota and those that the zoologists Jim Shaw and Tracy Carter followed at Wichita Mountains Wildlife Refuge in Oklahoma stayed with mom a year or even two years longer than those on Catalina.

I'm not certain why daughters separated from mothers earlier on Catalina, but I have a hypothesis that supposes they're trading the benefits of staying with mom against the benefits of maximizing good nutrition. Staying close to mom is safer and less socially stressful. When some ten-month-old orphan calves were put with a herd of strangers, they were harassed more than were same-aged calves whose mothers were present—most of the harassment came from animals one year older. Yet staying close to mom also has a cost: spending time in places or groups where a growing bison finds it harder to get its fill of needed grass. The more mom's needs and capabilities differ from her daughter's, the greater the cost. Scarce food and far-apart water sources would exacerbate such a cost. Food was much scarcer on Catalina. The whole island had once been severely overgrazed by cattle and goats. The range was being cared for and was recovering, but forage was still scarce there. The adult bison were small and, as noted in chapter 2, only about 30 percent of the mature cows had a calf each year. At Wind Cave and the Wichita Mountains forage was much better, bison were bigger, and 75 percent or more of the cows calved each year. Daughter bison may be programmed to detach themselves under nutritional stress: they may be designed so that hunger and thirst are stronger forces than the attraction to mom. These are the costs and benefits a calf must weigh, and their difference in different areas may determine when daughters leave mothers.

While daughters must choose when to end the relationship, mothers must choose where to begin it—where to give birth. She and the calf need some way to ensure that each develops a relationship with the right other individual—smells the right smell, hears the right voice. For though they have been intensely connected physiologically—sharing one body and one blood supply—socially they are complete strangers. They must create the relationship quickly and surely, before the precocious calf becomes so mobile that it mingles with others that look and sound very like it.

Cows can make sure of that privacy by being alone when they give birth. On Catalina Island most cows gave birth in solitude, usually among bushes, scrub oak, ceanothus, or coast live oak trees. There they were hard to see. They had both privacy and shelter from prying eyes—they had cover. But on the National Bison Range most cows stayed in their herd to give birth. There were exceptions to both rules, but the general patterns showed a striking distinction between the two places. This difference must have a cause, and my reasoning centers around the trade-offs between privacy and predation.

Wolves hunt mostly by sight. If wolves are your worry, then being out of their sight is best. If there are bushes or trees you can hide among them. But if the tallest plants are ankle-high grasses, the only thing big enough to hide behind is another bison—or better still, a bunch of bison. On the National Bison Range I was watching cows give birth on a grassland; on Catalina Island they were giving birth in a coastal scrub community.

Wherever the relationship begins, it eventually ends. The relationship is strong for several months, but after only one week the calf's independence strengthens. It leaves the mother more often and stays away longer, usually in the company of several calves. We see the most detail about this stage of a calf's development in Wendy Green's data on thirteen cows and their daughters. As the daughters entered their third month, their behavior changed sharply. They suckled less, rested less, and grazed more, usually with their mothers.

Keeping in touch with mother is a matter of life and death for the calf, and keeping in touch with her calf—her stake in the next generation—is almost as important to mother. Gestation in bison is nine months, so the breeding season starts just as most calves enter their fourth month. Some of the mothers Wendy was following got pregnant and some (mostly older cows) didn't. Pregnancy set up a classic conflict of interests, called *parent-offspring conflict*—the result of a competition between the best interests of the current calf and those of the future calf.

Selection has prepared a pregnant mother to prepare for another calf. Part of that preparation takes the form of ceasing to invest in her current calf so she can rebuild herself to deliver a healthy calf in just nine months. Thus the best deal for the cow is to invest less in this calf so she can invest more in the calf to come. She's equally related to both, and to her calves that may come even further in the future. That's not the best deal for the current calf. Mother's new calf is unlikely to be closer than a half sibling—sharing one-fourth of its genes. All things considered, its best deal is for mother to continue to invest in it no matter the cost to its future half siblings. The result is a classic conflict of genetic interest.

Pregnancy is crunch time. Now the cow has one calf at her udder and one in her womb. Her resources—her energy and her body's tissues—are finite, and she must divide them between her two calves. Wendy's pregnant cows' behavior toward the one at their udder changed sharply, unlike the behavior of the mothers that didn't get pregnant. Through the next three months (until the calf was six months old), pregnant mothers were more than twice as aggressive toward their calves when they nursed and attempted to nurse. From six months on, the differences were even more dramatic. Pregnant mothers nursed only 10 percent as much as nonpregnant mothers, and spent more than ten times as much time grazing while the calf nursed.

But the calf isn't just pushed away, it also walks away—choosing to spend more and more time with age-mates and getting more and more of its nutrition from the grasses it grazes. Sons become mere acquaintances first, but daughters eventually do too, when they have their first calf, if not before. Each is pursuing the path that will maximize the number of its genes in the next generation. They have moved on. The bond has served its purpose and dissolved. Daughters will experience its intensity again when they become mothers, but sons will never again experience a relationship anything like it.

PART TWO

The Machinery of a Bison's Life

An animal is a set of mechanisms—a collection of physical and chemical phenomena. Oxygen comes and carbon dioxide goes. Nerve impulses stream from the eyes and ears to the brain and from there to the muscles. The muscles contract or relax. There is so much to tell about mechanisms. Much too much to include here. But there is space to touch on a little of the physics and chemistry of bison life. I've chosen three aspects: movement, digestion, and body temperature maintenance.

CHAPTER 4   Bison Athletics

Even though they're imaginary, we can't imagine Elsie the contented cow or Ferdinand the flower-loving bull winning a footrace with Silver, the Lone Ranger's horse—and they couldn't. But Harvey Wallbanger, a flesh-and-blood buffalo, regularly showed his heels to racehorses in the 440-yard dash. To be sure, these humiliated horses were not the fastest ever to go to the starting gate—in fact, many were among the slowest in America's racing stables. Still, they were racehorses, while Harvey was basically an off-the-rack buffalo: the one who happened to be handy when a shrewd cowboy decided to go into the buffalo-racing business.

Harvey's triumph would not have surprised the Sioux, Crow, Blackfeet, Comanches, and Cheyennes who hunted buffalo from horseback for nearly two centuries. While most of their horses could overtake one buffalo, only a few could overtake several buffalo in one chase. A buffalo's skinny rump and long front legs give it a long-enduring stride—a good match for a coursing predator like the wolf. It is an animal faster than, well, some speeding racehorses, and able to leap tall road cuts at a single bound.

Grazing buffalo show no signs of Harvey Wallbanger's athleticism. They plod in short steps from one mouthful of grass to the next. When they move to water it's at a faster but still patient and economical walk. They lie down and get up with deep sighs and a cautious folding and unfolding of their legs that suggests the outcome is in doubt. Yet they are capable, at any second, of a memorable athletic moment.

I'm watching a mature bull standing alone on a dirt road on the National Bison Range. He's the only buffalo around, and I have set up my movie camera, so I'm watching him through the viewfinder—finger on the shutter button—wishing, as a man with a movie camera will, that the subject would do something footage-worthy. He stands broadside to the road's line of travel, his front feet at the bottom of the cutbank where the road is in a trough cut through a low hill to ease the grade. His right horn slips into the cutbank and cuts a horizontal groove. He glances up to the top of the cutbank, six feet above the road, cuts another grove with his left horn, glances up again, then—without seeming to gather himself—leaps to the top of the cutbank, lands upright on all four feet, and calmly surveys his new view. My finger is still on the shutter button, and I still haven't pressed it. I've just seen 2,000 pounds of buffalo do a standing high jump of six feet. My breath is quick and a little shaky, but the bull is perfectly calm. After standing for a minute he plods off. No high fives, or twos for that matter, but his patient, confident amble seemed an understated celebration of its own—"Not bad for a big bull with a skinny butt, eh?"

At a cattle guard, the road through a fence is open to vehicles but not to livestock. You make a cattle guard by cutting a hole in the fence as wide as the road, digging a shallow pit the width of the road and (usually) eight feet long. Then you cover the pit with parallel steel bars three or four inches wide, an inch or so thick, and as long as the road is wide. Set them on edge about six inches apart, perpendicular to the alignment of the road. Wheels easily roll across it, but cattle, faced with the choice of a long broad jump or a walk on a surface their feet might slip through, choose neither and so stay on one side of the fence.

Management installed some cattle guards on the National Bison Range. They were working fine for buffalo cows and calves, but not very well for bulls. Bulls were getting past them somehow, and one day I saw how. A bull walked calmly up the road to a cattle guard, stood placidly on one side of it, then hopped—no other word would really describe it—across, landing on all four feet on the other side. This hop had to be long enough to deposit his hind feet on the far side of the cattle guard, so he cleared the width of the cattle guard plus the distance from his front feet to his rear feet, say another six feet, for a total of fourteen feet. A very impressive standing broad jump. Well, at least I was impressed. If the bull was impressed, it didn't show. He stood where he had landed for a quiet moment, then, with an air of "been here, done this," cropped a mouthful of grass from the side of the road and walked on—patiently and efficiently.

National Bison Range personnel countered the buffalo hop strategy by placing two cattle guards end to end. Now a bull would have to hop sixteen plus six feet; and so far as I know, none ever did. But buffalo bulldom had not exhausted the arrows in its quiver. One day in breeding season I pulled my ancient Jeep to the side of the road, just after passing through one of these amplified cattle guards, and sat looking at the herd ahead of me. From behind came a distinct pinging, as though someone were tapping something metal with a piece of wood. A bull's image filled most of my rearview mirror. He was in the middle of the cattle guard, placing his feet delicately, cautiously, one at a time but still confidently, so they were centered on the narrow bars of the cattle guard as he walked across with all the poise, and a good bit of the daring, of a man on a tightrope. I would not have been any more astonished if he had also been singing "Tiptoe through the Tulips" in a friendly falsetto. When he reached solid ground he walked past me and joined the herd, leaving me to ponder the demonstration of footwork finesse I had just witnessed and somehow make it fit with the demonstrations of brute power I had also seen.

For while buffalo leaps and sprints are spectacular, walking is the athletic talent that brings the animals to food and water day after day. Bison are roamers. Even in the confined spaces where they live today, they will travel ten or twelve miles overnight. On the Great Plains they may have traveled hundreds of miles from season to season—perhaps searching for a better place to spend the winter, or for a location with fewer human beings. They surely gained something from each step of those journeys, but (and here is where a physical feat is required) to be profitable each had to gain them enough to offset its costs. And the cost is high; bulls weigh about a ton. When a vehicle that size is fueled with blades of grass, every blade has to count. So the athletic challenge becomes like one of those competitions to see how far a vehicle can travel on a gallon of gas. It's all about efficiency—getting the most out of every drop of gas or blade of grass.

Why is it that an animal that runs so fast walks so slow? It's all about energy. Buffalo, and just about everything else that walks, set a pace that matches the natural period of the pendulum constituted by its leg. A buffalo's leg, like yours and mine, swings forward and back as the animal walks, so it's a pendulum. The key point is that every pendulum has a most efficient swinging speed, the speed at which it uses the least energy to complete a cycle: that's its natural period. When you walk at your legs' natural period, a good bit of the energy that moves a foot forward at each step comes from your legs' pendulum swing; in this way, each time a foot swings forward it's recovering part of the energy invested in the previous step. At its natural period pace a buffalo, or any other four-footed beast, can recover 35 to 50 percent of the energy put into each stride. But when it comes to walking, two legs *are* better. Bipedal striders (creatures like ostriches and us) recover more, maybe as much as 70 percent, of each stride's energy just by walking naturally.

How fast is that? A pendulum's natural period is determined by its length, but something a little tricky happens here. Suppose you found a bison whose hind legs happened to be the same length as yours and you walked beside it (at a safe distance), matching it stride for stride as it walked to water. Strange as it seems, you would not both be walking most efficiently. Your legs and his are the same length as legs, but not as pendulums—the "length" of a pendulum is determined by the distribution of its weight. The closer a pendulum's weight is to the place where it pivots, the shorter the pendulum and the shorter its natural period. A buffalo's legs are heavier at the top and skinnier at the bottom than yours, so its natural period is faster. You'd have to hurry your pace to keep up. How much you'd hurry would depend partly on your choice of shoes that day. Featherweight footwear would make your leg a significantly shorter, faster pendulum than would a five-pound pair of waffle-soled, insulated mountain boots. Come to think of it, moccasins would be about the best choice you could make.

But you can bet your best moccasins that Harvey Wallbanger didn't walk away from those racehorses. Both parties were galloping flat out for a quarter of a mile, and both could gallop—a little more slowly, to be sure—for miles and miles, as most hoofed animals can. How do they get the energy? By conserving it. This illustrates not the pendulum effect, since the bison's legs are moving much faster than their natural period, but more a pogo stick effect. As their feet land, they store the force of gravity in tendons and ligaments threaded the long way around the joints in their legs. When their legs flex with gravity, those ligaments and tendons stretch like the spring on a screen door, and that energy is recovered as the leg straightens for the next step. Imagine a door spring attached at one end to the back of your knee, stretched down the back of your leg over your heel along the sole of your foot, and attached at its other end to the bottom of your toes. If you ran on your toes and flexed your ankle with every step the spring would store energy as you landed on your foot and return it as you strode ahead. That's one of several springs in a hoofed animal's leg and foot—sort of an elaborate pogo stick. (But it's also different from a pogo stick, which stores energy by compressing a coil spring; a tendon stores energy by stretching.) Sheep can recover about 30 percent of their running energy this way, and camels may recover 50 percent. Buffalo fall somewhere in that range.

We humans don't have the feet for this feat. Bison are always on their toes: that joint about a third of the way up their leg isn't a backward knee but the heel of their foot, and the tendon from their real knee to their toes is long and stretchy.

THE QUICK AND THE DEAD

Bison athleticism isn't all track-and-field events and efficiency contests. They fight, too (at least the bulls do), and power alone won't win a fight or even get a contender out of one alive. A bull has to twist and turn—quickly enough to protect his own flanks, quickly enough to get a horn into his opponent's flank. Selection is intense. Bulls are wounded every breeding season, and in most years 5 or 6 percent of the mature bulls in any population die of their wounds.

So the bulls are built to be quick in battle. To protect their body with their head, they need to pivot around their front feet. They have a great form for that function: much of their weight is centered over their front legs—their diminutive rear end is balanced in part by their massive head and neck. And the weight of their head is partly suspended from a point above their shoulders. There, rays of bone a foot long (called *vertebral processes*) project up from their vertebrae and anchor a tendon that attaches to the rear of the skull. This efficiently supports the transfer of their head's weight to their front feet, on which they pirouette on the sod like a hockey player on ice.

When I started to study buffalo a colleague said, "But they're just humpbacked cows!" Both walk on four legs, eat grass, and ruminate, but cattle were selected in a competition to supply milk and meat to humans cheaply and safely. Bison were selected in a competition to produce more bison—a competition in which the better athlete wins. That competition honed their shape and substance from small haunches to high hump, from size and strength to agility, from speed to stamina.

People sometimes say the competition in which bulls won breeding rights selected the "best" bulls, suggesting they do everything well. In fact, their specialization has cost them. One day I watched three mature bulls and a young one chase a cow. For the first 200 yards the older bulls ran easily, their long hair flowing in the wind. They even tried to mount the cow at a full gallop. But when the cow and the young bull circled back, still at a full gallop after a long run, the mature bulls followed at a wobbly walk, tongues hanging out and sides heaving. They were just too big to keep up the pace the smaller animals set. Natural selection has compromised much else to focus the bulls on one goal—forcing other bulls aside at tending time. It is then—when the bull moves to his task, beard swaying and pantaloons bouncing, belly lifted in an arch as he bellows a challenge—that he is perfect. And he is magnificent.

CHAPTER 5

Digestion

Grass to Gas and Chips

When Buffalo Bill Cody took the Czarevitch on safari in what would become South Dakota, they rode through a sea of flesh at ground level, with more below ground and even some in the air. In just about every square mile, thousands of vertebrate hearts were beating every moment, sending hundreds of gallons of blood coursing through thousands of miles of blood vessels. All this fabulous, teeming mass of animal matter and its motion was built from, and powered by, a growing grassland.

A grassland's plants combine the energy from the sun with water and nutrients from the soil to grow and reproduce. These plants produce the stuff of life and growth for grass eaters. There are carbohydrates for energy and protein for growth and repairing body parts. Many a backyard lawn produces enough calories and nutrients, strictly speaking, to nourish its owner. But the process is not that simple, because plants fight back.

Plants have their own uses for the energy they capture from the sun, just as animals that are the prey of others have their own uses for the energy they capture from the plants. And like prey animals, plants have evolved ways to defend themselves. They can't run and they can't hide, but they can be unpalatable, difficult to chew, or indigestible. They can even be lethally toxic. Plants put a good bit of energy into their defenses. For example, many plants produce tannin, which does nothing for the plant but makes it difficult for the animal that eats it to digest proteins. Many plant eaters, including humans, have evolved a countermeasure— our saliva contains a molecule that binds with tannin and neutralizes it. The astringent taste of the neutralized tannin gives a sip of red wine its special flavor.

And so coevolution goes—measure, countermeasure, counter-countermeasure. It's a sort of biological arms race made inevitable by the fact of natural selection, and it has produced some elegant measures and countermeasures. Grass hasn't evolved tannin, but it stores most of its carbohydrates as cellulose, and until the cellulose has been digested the carbohydrates are not available. In turn, most of the Great Plains' big animals—bison, antelope, deer, and elk—counter cellulose with rumination, which turns grass into gas: figurative gas, as fuel to run their physical systems, and literal gas, as methane. Digesting anything is a strictly chemical matter of subjecting it to an enzyme that breaks certain molecular bonds, simple enough if you have the right enzyme. Put the food in your digestive tract, secrete the enzyme. Neither you nor I can secrete an enzyme that can digest cellulose. As a matter of fact, bison don't secrete such an enzyme either, but they rely on a method as good and in some ways better: they enlist colonies of bacteria.

ENZYME, ENZYME, WHO'S GOT THE ENZYME?

Reliance on bacteria is a wonderful strategy. Think of a thing, almost anything, that is or once was alive, and chances are good that some bacteria can digest it. That is to say, bacteria exist that have evolved an enzyme that can break the molecular bonds in such a way as to make the energy and nutrients available to those bacteria.

Bison didn't evolve an enzyme that can digest cellulose, but they did evolve a happy home for bacteria that have evolved such an enzyme. The front part of their stomach is segmented off by a fold (the rumen) in which newly swallowed food is kept for a while. The fold probably came before grass appeared on the earth. It created a special place where food, as soon as it was swallowed, could be detoxified—the plant's chemical defenses neutralized—by some toxin-eating bacteria living there. But now it serves as a place where some very helpful bacteria put their enzymes to work digesting the cellulose.

The ruminants have enlisted a powerful ally in their arms race with grass. A key to success in any arms race is how quickly measures can be developed to counter an opponent's previous development. The advantage therefore goes to the party that can evolve faster. Ruminants don't evolve faster than grasses, but bacteria, with dozens of generations in a

COW PORTRAIT

This cow has some soil caught in her eyelashes. Freezing water cannot accumulate in the extremely short hair surrounding her eyes. Domestic cattle are sometimes blinded by ice accumulating in the longer hair that surrounds their eyes.

BULL PORTRAIT

The thick cushion of head hair may make a mature bull seem larger to competitors. It certainly absorbs some of the shock of battle. The base of a cow's horn is about the size of her eye. The base of a bull's horn, even in a young bull, is much larger than his eye.

BREEDING HERD

This breeding-season herd is on the National Bison
Range in western Montana. The mature bulls join the
cow groups from mid-July to mid-August and
compete to tend estrous cows.

The bull on the left threatens by standing stiffly broadside. He is also bellowing, with the tip of his tongue touching the roof of his mouth. The bull on the right turning his head away is signaling submission.

THREAT WALLOWING

Challenging bison often wallow between bouts of bellows. Sometimes they use an established wallow; sometimes they tear up sod with their horns to make a new one. In either case, they often urinate into the wallow before rolling in it.

POST–THREAT WALLOWING

After wallowing less than half a minute, the challenging bulls rise to confront each other again.

HEAD-ON THREAT AND SUBMISSION

Two bulls were approaching each other after
wallowing, when the bull on the right yielded by
turning his head to the side.

NOD THREAT

The bulls approach closely with their heads to the
side and intermittently swing their heads up and
down in matched movements. If there is an attack, it
comes during the head-low phase of the cycle.

FIGHTING

In a herd with a natural sex ratio, about 5 to 10 percent
of challenges lead to fights. In a breeding herd there
can be dozens of fights on some days. The other herd
members become habituated to the drama.

·

HEAD-TO-HEAD CLASH

Most fights are head-to-head clashes followed by
head-to-head pushing. To point their horns forward,
bison bulls must drop their heads very low.

FLANK ATTACK

Most contact is head to head, but the bulls
also circle or push past the opponent's defense
and try to gore an exposed flank.

GORED BULL

This bull was gored about ten days before being
photographed, and the wound has become infected.
Such infected wounds often kill bulls, but this bull
recovered and was healthy two months later. A third
of mature bulls have at least one rib that has been
broken and has healed.

HEAD HAIR SHEARED BY FIGHTING

When bulls fight, the hair on their foreheads and
between their horns is often caught between the
horns, sheared off as the horns slide past each other,
and tossed into the air. Such a lock of hair
is in the air above these bulls.

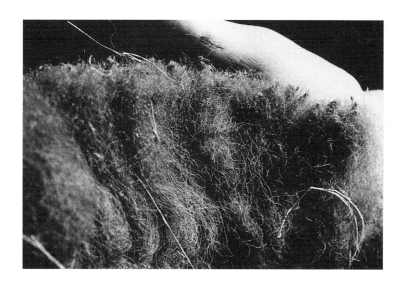

LOCK OF HEAD HAIR

This lock of hair was sheared off in a fight.
The shearing action of the horns has given
the upper edge a straight line.

DISPLAY HAIR

Left Tip, named for his left horn's flattened tip, was
photographed on July 22, 1972, with his head hair, beard,
and front leg pantaloons intact.

.

DISPLAY HAIR LOSS

Left Tip was photographed on October 3, 1972. Most of the
hair between his horns had been sheared off in fights, but his
beard and pantaloons were still full at the end of the
breeding and fighting season. They then molted.

LIP-CURL

The lip-curl is technically called *flehmen*. Bulls thrust their muzzle into a cow's urine, then lip-curl. The bull's tongue flutters up and down rapidly, probably transporting the urine to his vomernasal organ in the roof of his mouth. Sensitive to the metabolic products of a cow's reproductive hormonal cycles, that organ enables the bull to predict when she will become receptive.

TENDING

During the midsummer breeding season, individual
bulls accompany particular cows—usually cows
that will soon breed. In a given day all the bulls
in a herd compete to tend the cows most
likely to breed that day.

RUNNING COW

Shortly before they become receptive, most cows
break away from the tending bull and run through
the herd a few times. These runs attract the attention
and pursuit of other bulls, thus inciting
competition for tending rights.

COW MOUNTING BULL

Tending bulls sometimes become so preoccupied with their rivals that they don't detect the cow's readiness to breed. A neglected cow may mount the bull. This usually gives him the idea.

COPULATION

Penetration: at this point cows often begin to walk
or run, but some stand—as this one did.

.

EJACULATION

The bull's hind feet are lifted clear of the ground by
the contraction of his abdominal muscles.

WINTER FORAGING

These cows and calves are feeding in deep snow in Yellowstone Park. Bison reach the snow-covered grass by sweeping the snow away with their muzzles.

SNOW CRATER

A cow feeding on grass at the bottom of a crater
that she has cleared in the snow. A dominant cow
could displace her and get the food benefit without
any cost (paid in the effort to clear the snow).

WEEKS-OLD CALF NURSING

Bison calves are precocial—on their feet and
able to walk and nurse less than ten minutes after
being born. The pulse of oxytocin that facilitates
birth for the cow also facilitates the mother's
bonding with her calf.

MOTHER AND CALF

The relationship between mother and calf is
by far the strongest and longest in a buffalo's life.
Bull calves stay with their mothers until they are
weaned at about six months old. In some habitats,
heifer calves stay with their mothers until
they become mothers themselves.

GROOMING

Bison groom their hair regularly, primarily with their
tongue, but they use their horns and hooves
on spots their tongue can't reach.

WALLOWING

All bison wallow several times a day in summer,
filling their hair with dust. The dust probably
discourages insects and may reduce the
bison's body temperature.

Bison cool their bodies by evaporating water from their lungs. On hot summer days they need a lot of water. On Catalina Island, cows go to water twice a day, drinking four to six gallons at a time. Bulls there drink once a day in the summer. I was unable to measure the amount consumed by the average bull.

BUFFALO BIRDS

These birds, now usually called "cowbirds," feed on
the insects that flee the mouths and hooves of grazing
buffalo. Since bison are nomadic, a buffalo bird's food
supply may move several miles from one day to the next.
Buffalo birds are parasitic brooders, laying their eggs in
other birds' nests. Thus they too are free to roam.

PRONGHORN DOES AND FAWNS

Female pronghorn and their fawns move about in
small groups, spending most of their time in the parts
of the grassland with the greatest number of broad-
leafed annual plants. When and where pronghorn are
territorial, master bucks set up territories that
include these preferred microhabitats.

PRONGHORN MALE
IN MIDSUMMER

Although pronghorn have
true horns (not antlers), the
horn sheath is shed every fall,
revealing a small bony core
around which a new sheath
forms in the spring. A scent
gland lies under the dark
cheek patch.

PRONGHORN MALE MARKING

A territorial male wets the top
of a tall plant in his mouth, then
coats it with a secretion from
his cheek patch gland.

WHITE BUFFALO

A very rare white buffalo calf, born on the National Bison
Range in 1937, is pictured with my grandfather. Then
superintendent of the National Bison Range, he had bred
an almost completely white bull back to that bull's mother.
This was their offspring. Granddad appears to be smoking
a celebratory cigar. White coats would absorb less of the
sun's energy in winter, making cold weather more
costly for white buffalo than for dark ones.
(Photo credit: my mother, Joyce Norton Lott.)

PLUGGED PRAIRIE DOG BURROW

The burrows of black-tailed prairie dogs have one high entrance and
one at ground level; the differential creates wind-powered ventilation. When
rattlesnakes enter the burrow the prairie dogs sometimes use soil from the
mound to plug the opening, as Debra Shier saw a male do here weeks ago. He
then plugged a nearby ground-level opening. The prairie dogs usually reopen
the burrow after a day or two, which allows the snake to escape.

·

SURVIVORS

A litter of young prairie dogs has reached midsummer
without succumbing to predators or infanticidal relatives.

WILD BULL RESISTING ROUNDUP

The bull pictured is reacting to men trying to move
him through a corral on the National Bison Range
during the annual roundup. Handling wild bison is
difficult, dangerous, and expensive. Domestication
will select for more tractable—hence less wild—
animals. This bull's behavior will be tolerated in this
publicly owned herd, but a rancher would be
compelled to shoot him.

SADDLED STEER ON CATALINA ISLAND

Some buffalo, usually castrated males, have been trained to
carry a rider or pull a cart. They remain dangerous.

.

HORSES

Reintroduced to North America by the Spanish after a 10,000-year
absence, horses transformed the buffalo's world. They multiplied
the take by Native American hunters, supported European hunting,
and competed for grass against both managed and wild herds.

RANCH HERD

This herd was photographed on Vermejo Ranch in
northeastern New Mexico. Bison are better suited to
western grasslands than are cattle, requiring less care
and doing less damage. More than 95 percent of bison
in North America are privately owned and live on
ranches. There selective breeding will produce a
domesticated form of bison with little wildness left.

day, evolve faster than just about anything else. By enlisting bacteria in their cause, ruminants shifted from being slow to being blindingly fast in the evolution of countermeasures. The rumen becomes an arena in which different kinds of bacteria compete, enzyme against enzyme; the producer of the most efficient enzyme succeeds in expanding its population at the cost of the populations of its competitors.

Bacterial digestion of plant materials is a common process with a common name—fermentation. And although bacteria are simple as life forms go, their chemistry is complex, and fermentation produces a complex outcome. They not only change cellulose to usable carbohydrates but also produce volatile fatty acids. Both are concentrated energy in a form bison can use. They are the gas from grass that makes the animal's heart beat and its feet move. Like all living things, these fermentation bacteria have waste products, which include alcohol. It's a sobering fact that 12 or 13 percent of a bottle of Dom Perignon Champagne is bacteria pee.

For the bison a percentage of bacteria waste products that high would be a calamity, because it would mean that the bacteria were themselves using the energy their enzymes were releasing. Here we come to one of those built-in conflicts of interest that are part of nearly all relationships. Up to the point of converting cellulose to usable carbos and fatty acids, bison and bacteria have the same goals and collaborate. But now each has its own uses for the energy the enzymes have released—now they compete. It has to be a restrained competition, because both would starve if either were to get all the energy. Yet within the rumen, subtly different lines of bacteria must be striving to win a bigger share of the goodies. The bacteria-on-bacteria competition takes place in a friendly environment— the bison's rumen—so the bison, being the environment as well as collaborator and competitor, has leverage that makes up for its slower evolutionary rate.

When I was a soldier nearly fifty years ago, the cooks had a motto, "Just feed 'em—don't fatten 'em," which they enforced by rejecting requests for seconds. Bison don't limit their bacteria's food, but they severely limit their oxygen and their time. Some bacteria—anaerobic bacteria—can function without oxygen, but they function slowly. In time they would use up the energy they have released from the grass, so bison don't give them time. They move the grass on out of the rumen after two or three days, pushing the partially digested food, and some of the bacteria that digested it to that point, further on. In the stomach, their own enzymes finish the job on the plants and digest some of the helpful bacteria as well. Timing is everything in this matter, and the ruminants have the timing down so well that they get about 90 percent of the energy for themselves.

The mechanics of rumination are a bit inelegant. The ruminant assists the process by chewing its food after swallowing it—the bison brings up fist-sized wads of partially digested grass (cuds) from its rumen and chews them while lying at rest. If Buffalo Bill and the Czarevitch had sat in their saddles and contemplated a resting bison they'd have seen this; bison spend hours every day doing it. Although cud chewing—ruminating—gives them a faraway-focused, meditative, serene sort of look, it isn't an elegant activity. But as an adaptation, rumination is elegance itself. It is an adaptation that uses the process of adaptation—a sort of meta-adaptation, if there is such a thing—and it is beautiful to contemplate. It's so sophisticated that neither bison nor biologists would be likely to think of it, yet it was achieved by the perfectly purposeless, aimless, and automatic process of natural selection.

BUFFALO CHIPS

The immediate products of bison digestion are heat, energy, and tissue maintenance for the digesters' bodies. The final end products of bison digestion are buffalo calves and buffalo chips—a good many more chips than calves. People living on the plains burned buffalo chips as a cooking fuel. When they thought of chips, they thought of heat. When we in the computer age think of chips we think of information. A biologist who really knows his or her . . . well, chips can extract an enormous amount of information from one.

To appreciate a buffalo chip, you must ignore its rather unfortunate looks and any lingering odor, and think instead of where it's been. It has gone entirely through the animal from front to back, and by a very circuitous route. And every inch of the way, it has picked up traces of what is happening around it, while retaining its own identity. Think of it this way: if a computer memory chip traveled a similar electronic route, and had bits of data added to it at every step, it would emerge loaded with information. We couldn't wait to install it and read it. Small wonder, then, that biologists constantly upgrade their capacity to read those memory chips that digestion produces. Some years ago we figured out how to read in them a record of what the bison has eaten and what parasites live in its intestines.

Analyzing parasites is the most straightforward part of this study—just look and count. Figuring out what bison are eating is also straightforward, though rather more time-consuming. The cell walls of plants are pretty distinctive and pretty tough. You pick through the chips and locate cell walls. You compare the cell walls to the cell walls of plants from your reference collection—representatives of each plant that grows where the bison are feeding. The reference collection is needed to set some boundaries on your search for matches. And that's it. You can find out how many of the plants in its habitat a bison feeds on and, by comparing the percentage appearing in the chips with the percentage in the field, estimate which plants the bison favors.

Recently we've learned to track hormones from their residues in feces. Tracking hormones is easier than determining diet, but a little more high-tech. Progesterone shows up in feces. With a bit of biochemical abracadabra called *enzymeimmunoassay* you can discover how much a given (or taken) fecal sample contains. After you spend some more time, money, and enzymeimmunoassays calibrating the chips from bison feeding in a particular place, you can tell from her feces if a cow is in estrus.

In their passage chips also pick up bison cells that contain the individual's complete genome. It is possible that they could reveal not only the individual's identity but perhaps the identity of its parents as well. So science will just keep chipping away at the secrets in the belly of the beast.

What we can determine now is probably just the beginning. Everything that happens inside an animal produces a by-product that will reveal something about that process, and nearly all those by-products (and products) find their way into the lower end of the digestive tract and into the chips assembled there from the materials that have made the passage. There's probably at least one other way to get each bit of that information, but few other ways are as humane and efficient as chip analysis. No need to subdue the buffalo with a tranquilizing dart—and no worries that the hormone levels in the blood sample reflect short-term peaks or bottoms caused by the trauma of the sampling. Little wonder, then, that when the chips are down, the biologist's spirits are up. The investigator that at first seems a figure of fun, a dedicated pooper-scooper, is really the very model of a modern-day mammalogist.

CHAPTER 6

Temperature Control

Big Medicine was the first white buffalo born in the twentieth century. His mother lived on the National Bison Range, where she gave birth to him in May 1933. My mother lived on the National Bison Range too, and she gave birth to me in December 1933. My maternal grandfather was the Bison Range superintendent then, and my dad worked for him. I won't say Granddad liked Big Medicine better than me, but the family photo collection contains many more snaps of Granddad and Big Medicine than of Granddad and me. Ah, well, a lot of us are born into somebody's shadow, and while I never outgrew Big Medicine's I have outlasted it by several decades. My family members weren't the only photographers besotted with Big Medicine—he was surely the most photographed buffalo that ever lived.

Big Medicine was not a true albino. His eyes were pigmented and he had dark hair between his horns. But the rest of his hair was white. I've had a lot of years to think about Big Medicine, and a frequent thought is this: Why aren't all buffalo white?

White is a good summer color. It reflects the sun's heat, and that heat is a fact of life for bison. They evolved in the grasslands, where summer sunshine is plentiful and shade is nonexistent; bison spend most of the long summer days absorbing the sun's heat into their dark coats. White is also a good winter color. Some rabbits turn white in the winter to blend with the snow, and so do some of the weasels that stalk them. Wolves hunt bison all winter long. Wouldn't blending with the background be a good idea? Rocky Mountain goats and Dall sheep—wolf prey on both—live around snow and wear white coats year round.

But not only are white buffalo rare, they don't seem to do well. Far from going forth and multiplying, they dwindle and disappear. Granddad tried to promote them. With the directness of the stockman and veterinarian that he was, he arranged for Big Medicine to breed his mother as soon as he was old enough. They had a son who was a true albino. The nation's capital called and he was sent to the National Zoo in Washington, D.C. There he either ripped his leg in a fence and died of the resulting infection, as a zookeeper described to my brother, or swallowed some baling wire along with the baled hay he was being fed, as another story has it, and died too young to reproduce—zoo personnel have told both tales. Granddad intended to try for more, but the bureaucrats above him—wisely, I think—decreed that the United States Fish and Wildlife Service was not in the business of producing freaks of nature.

But nature has produced some in Alaska. When the state wanted to start several herds there, it got its stock from the National Bison Range. Big Medicine's gene was revealed in the coats of several calves born in Alaska, but none of them lived long enough to breed. Their early deaths may tell us something important about why white buffalo are so rare. The answer probably lies in winter. Someone once said, "Powerful though the sun's abundance is, its scarcity is even more powerful." Bison seldom if ever die of heat, but they often die of cold. The dark coat that makes the sun a nuisance in summer may be a lifesaver in winter. Bison evolved in really terrible winters; and even now, especially severe winters kill many of the old and the young. The sun is low and the days are short, but every calorie of heat absorbed from the sun is a calorie the bison does not have to manufacture from the scarce forage—forage that must be won by sweeping the snow from each bite with that heavy head—or drawn from its precious cache of calories in the form of stored fat.

Like deer and elk, bison cut their energy output by losing their appetite. They eat less and produce less heat—and not just because food is scarcer in winter. Even when they can have all they want from full feeding troughs in an experimenter's corral, they eat 30 percent less food and produce 30 percent less heat in February and March than in April and May. The aphagia cuts down on low-profit energy expended in searching for food in a season when it would be scarce, but to accumulate the sun's energy they still must depend heavily on that dark hair.

But there's more to hair than color; it also offers insulation. A high-tech way to census large, warm-blooded animals in winter is to fly with an infrared device—one that "sees" temperature. Wes Olson is the chief ranger at Elk Island Park in Alberta. He added an infrared detector to his elk, moose, and bison winter census a few years ago. The first flight was illuminating. There was a thick layer of fresh snow and the elk and moose stood out clearly against it, but hardly any bison showed up. There *were* a lot of warm crescents that made no sense. The eyes revealed what the infrared scope could not—the crescents were the bellies of reclining bison, the only part of them not covered by a thick layer of still-unmelted snow. The skin-side layer of the bison's coat was at body temperature, while the outside layer—only millimeters away—was below freezing.

A buffalo robe was a possession prized by humans dwelling on the North American plains. It can keep you warm in a terrible storm. Cattle have replaced the buffalo, but there isn't much point in bundling up in a cattle robe. The buffalo robe's superiority is quite straightforward—a square inch of buffalo skin has ten times as many hairs growing from it as a square inch of cow skin. The difference, when temperatures and fat stores are low, is the difference between life and death.

But when summer arrives there is a price to pay. The North American plains are a place of extreme heat as well as extreme cold. It's a bit like jumping from the deep freeze to the frying pan, and the challenge in summer is keeping cool. The first thing the bison do is shed their winter coats. The long, twisted, almost woolly hair of winter molts, and from the front shoulder back a sleek coat of short hair is revealed. It insulates only a little, allowing the nearly ceaseless wind of North America's grassland to blow away body heat. That surely helps; but still, the sun is hot, their dark hair absorbs its heat, and they also produce heat as they ferment their food and move around. If they couldn't get rid of the heat they generate and the heat they accumulate, they would soon be walking pot roasts.

Lots of animals get rid of heat by evaporating water. Humans sweat, and when the sweat evaporates from our skin the process consumes heat, which is taken from our skin. The process has created a niche—our armpits—for the deodorant industry. Bison don't sweat, but they breathe, and lots of animals, again including humans, lose heat by evaporating water in their lungs. It was midsummer when Doug Propst first laid eyes on the Catalina Island bison herd he had been hired to manage. Doug was a Colorado cattleman, born, bred, and college-educated there. He told me that when he saw the first group of bison lying in the summer sun he was appalled. Their respiration rate (about forty breaths per minute) was consistent with only one thing. The whole herd had pneumonia! Imagine his relief on discovering that they were only keeping their cool.

Evaporative cooling works well, but it has one major cost. You have to go through a lot of water in order that a lot of water can go through you. That internal supply must be replenished regularly. Surely the bison can do something to reduce the number of trips to water. Of course they could find some shade, but they don't seem much inclined to. It's astonishingly common to see them lying in the hot sun only twenty feet from dense shade. Perhaps their roots in the grassland where they evolved are most exposed here. There were no trees and no shade where they came from, so any inclination to search for it would have been a disadvantage.

But grasslands at least have soil, and here we may be onto something. Bison wallow in the summer, especially during the middle of the day. Wallowing puts soil into and onto their coat. They can work so much nice, dry, powdery soil into their coat that as they walk away from the wallow it cascades down, jarred loose by each step. Like most old-timey bison watchers, I have always thought they were wallowing to make their hair a lousy place for lice and other parasites. I still think that's likely, but wallowing may also lower their heat load. Elephants have a heat problem much like bison have, and we know that a good coat of dirt is one of their solutions. We don't know that about bison—yet, but the odds are good that it also works for them.

PART THREE

*Whence They Came Forth,
and How Much They Multiplied*

Every species has a history, and that history is a part of the species as much as an individual's history is a part of the individual. It both creates and limits the species' possibilities. Chapter 7, on ancestors and relatives, traces the bison lineage, especially in North America, and establishes the place of the modern species Bison bison, *which emerged only about 5,000 years ago.* Bison bison *became one of the most abundant large land animals of all time. Chapter 8 addresses the question of how abundant it was before the European invasion of North America.*

Ancestors and Relatives

Perhaps he fled the lions, his flying hooves hammering the grassy turf as he dashed past a herd of woolly mammoths or scattered a herd of Saiga antelope. Perhaps, being a prime bull at eight or nine years old, he faced and fought them. We don't know the details, but we do know the lions prevailed: in a matter of minutes there was one less buffalo alive in Alaska. His killers were American lions, similar enough to those in Africa today that—though they were a bit bigger—the casual observer probably wouldn't distinguish them. Very likely they hunted and killed in the same way.

We know the essentials about this violent death—who was killed, who killed him, and how and when and why (the lions were hungry) he was killed. There was a witness, the dead bull himself, and though he was silenced 36,000 years ago, he has testified through the forensic skills of the paleontologist Dale Guthrie. He is not one of the anonymous dead; he has a name: Blue Babe. He is here to tell his story because after the lions killed him and made a meal of his hump, a mud slide buried him. The mud froze, and the mud and Blue Babe remained frozen until a gold miner washed away the mud and revealed the mummy one July day in 1979.

Copper precipitation had given Blue Babe's hide a blue tint and his hump had gone to fill the lions' stomachs, but the rest of him was remarkably well preserved. The tooth and claw marks in his hide were still so clear that Dale could take an American lion's skull, place its canine teeth on the marks left by the killer's canines, and see a perfect match. Even the flesh was so well preserved that when the corpse had yielded all its secrets Dale and his colleagues made an acceptable stew with a bit of the meat. Blue Babe had died in good condition, probably in early winter.

Blue Babe was a *Bison priscus*—at least two phylogenetic steps back from today's *Bison bison*. He was very like his ancestors that came to North America from Siberia. It wasn't a long journey. When the route was dry, bison could have walked from Siberia to Alaska in three or four days. During ice ages glaciers accumulate so much of the earth's water on the continents that the sea level falls below the northern part of the Bering Sea's bottom, which is then called "Beringia." When Beringia connects Siberia and Alaska above sea level, it is often called the "Bering Land Bridge." Calling it a bridge makes it sound narrow, but it's really a subcontinent, at times 600 miles wide—leaving plenty of room for gradual colonization and species dispersal in both directions.

The first bison to cross Beringia were not the first bison. Bison branched off from the primitive cow family line—*Leptobos*—about a million years ago. The first bison were small-bodied, small-horned, fast-moving residents of forest edges and meadows. Gradually the bison line became northern specialists, able to live where other cattle couldn't. They also became open grassland specialists. The Siberian steppe was a northern grassland, and bison became important members of that community. Bison taxonomists call this bison *B. priscus*. It was *B. priscus* that grazed its way from Siberia across Beringia to Alaska, perhaps as early as 600,000 years ago, perhaps as recently as 300,000 years ago. By then they had bigger bodies and bigger horns, and weren't quite as fleet of foot.

B. priscus stayed north, a widespread and abundant grazer on the ice-age region that Dale has called the "Mammoth Steppe." It includes not just Beringia but also eastern Siberia and northwestern Alaska. Winters there were cold, but grasses grew abundantly in the long days of summer and supported a large grazing community. The three most important grazers were woolly mammoths, horses, and bison. Mammoths and horses became a bit smaller in this environment. But bison, the only ruminant of the three, apparently took advantage of a ruminant's potential for rapid growth and became larger.

There is some disagreement about exactly what happened next. Bison fossils were first studied and classified at a time when taxonomists thought all members of a species were very similar to all others of the same age and sex class. Therefore, they saw modest differences between sets of bison skulls as signifying separate genetic lines and divided the bison on the Mammoth Steppe into several species. Some taxonomists still prefer that scenario. But we now know that many species are plastic. Their bodies and bones can differ in different places or at different times but still be from one species, able to produce healthy offspring from the mating of individuals of quite different sizes and shapes. Think of dogs. If we look at *B. priscus* with dogs in mind, the modest differences in skull shapes and sizes from place to place and from time to time fit easily into a picture of one plastic species persisting over several hundred thousand years. I side with those who favor this scenario.

B. priscus did produce some offshoots that were clearly different species. One was a giant that developed from a line that reached central North America. *B. latifrons* stood some 20 percent taller than modern bison and the bony cores of its horns spanned six feet. With the sheaths in place, the horns must have been nearly seven feet on the living animal; and *B. latifrons* must have weighed at least a third more than modern bison, putting bulls close to 3,000 pounds. *B. latifrons* may have appeared as early as 300,000 years ago. These bison lived all over the region that has become the United States, but they were most abundant in the West, along the eastern edge of the Rocky Mountains, in the Great Basin Area—Nevada, Utah, southern Idaho, and Arizona—and in California. A few years ago construction excavation unearthed a *B. latifrons* in downtown San Francisco. It had lived there 25,000 years ago, when San Francisco Bay was a wooded grassland valley with a river running through it.

The pace of bison evolution stepped up sharply about this time. The San Francisco *B. latifrons* was near the end of that line; no fossils younger than 22,000 years have been found. But *B. latifrons* probably gave rise to its successor—*B. antiquus*. *B. antiquus* lived in the same general regions, but was smaller bodied and much smaller horned. Its life as a species was short: it was born as *B. latifrons* died and then disappeared about 12,000 years later.

B. antiquus was replaced in much of its range by *B. occidentalis*, a smaller bison probably born of the *B. priscus* line still occupying the Mammoth Steppe. *B. occidentalis* was smaller and smaller horned than *B. priscus*; and unlike any bison before it, its horns pointed up, parallel to the plane of its face from nose to forehead, instead of pointing forward through that plane. For a short time *B. antiquus* and *B. occidentalis* shared the Great Plains, but *B. antiquus* soon faded away. *B. occidentalis* never expanded into all *B. antiquus*'s habitat—it never appeared in California, for example—but *B. antiquus* died out everywhere by about 10,000 years ago.

B. occidentalis had an even shorter life as a species. It prospered, not only filling most of the space *B. antiquus* left vacant but becoming much more numerous even as the other members of the grassland community— horses, camels, woolly mammoths, musk oxen—faded rapidly. Yet in only about 5,000 years it was replaced by today's still smaller descendant: *B. bison*. *B. bison* is the smallest bison that ever lived in North America and became by far the most abundant.

Bison changed little for at least a quarter of a million years, then changed a great deal during the last 10,000 to 12,000 years—*B. antiquus* died out; *B. occidentalis* appeared and then died out as *B. bison* appeared. Even more dramatic changes have recently overtaken other large North American mammals in the same period. While bison were reinventing themselves, the Pleistocene extinction, a radical downsizing of the number of large mammal species, swept North America. Horses, camels, mammoths, mastodons—after millions of years of evolution in North America—all vanished around 10,000 to 12,000 years ago. Their exit was followed, or accompanied, by that of a suite of large predators—the giant short-faced bear, the American lion, the dire wolf, several species of saber-toothed cats. Many bird species vanished at the same time.

There are several explanations for this great species loss. One theory argues that the trigger, in fact the fatal bullet, was the arrival of another large mammalian predator—humans. It's not certain just what manner of humans first came to North America or when they arrived. But it seems clear that North America was peopled by 12,000 years ago, and that those people hunted large mammals. There is a school of thought, led by the archaeologist Paul Martin, that lays the Pleistocene extinction at the moccasin-clad feet of these early hunters. In this Pleistocene overkill theory, Martin envisions a narrow band of human hunters extending from coast to coast and descending from north to south, killing virtually all members of most large prey species as they moved.

Most ecologists doubt that's the whole story. North America is a big continent, and it was inhabited by tens or perhaps hundreds of millions of large animals that were being born as well as dying. To drive even one abundant species to extinction would be a formidable task. To drive several large grassland species to extinction simultaneously would be a prodigious feat. That's one of the reasons many ecologists are skeptical about the Pleistocene overkill hypothesis. Another reason is that there were other powerful forces at work. Climate and habitat changed rapidly and dramatically as the last ice age ended 10,000 to 12,000 years ago. California, for example, no longer had summer rainfall. The grasses dependent on that rainfall died out, and buffalo, which need grasses that grow in the summer, dwindled away. The grasslands east of the Rocky Mountains became drier and were subject to periodic severe droughts. Habitat changes give some species the edge and give others the axe. We need not choose between climate and hunters. Both impinged on the animals and both must have had an effect. A plausible scenario is that climate and habitat changes depleted the populations and human hunters delivered the coup de grâce.

Human hunting wouldn't have affected all species equally, and it may have had a role in the recent rapid evolution of bison. *Bison antiquus* was still here when the human hunters came, but it seems to have disappeared within one or two thousand years of their arrival. One possible explanation is that *B. occidentalis* and its successor species, *B. bison*, dealt with the new hunters more effectively than did *B. antiquus*. Human hunters had a new killing technology—the bladed spear—that enabled them to deliver a mortal wound without coming within reach of mammoth tusks, horse hooves, or bison horns. The long, hooked shape and forward orientation of *B. antiquus*'s horns suggest it may often have dealt with predators by standing its ground and fighting. Stand and fight was a much riskier strategy with the new predator: those who lived by it were likely to die by the spear. The new technology made running away, as *B. bison* usually does and as *B. occidentalis* may have done, a better tactic.

At the same time, another predator's tactics also favored rapid retreat. As the other large grassland predators dwindled, gray wolves became the largest of those left on the plains. Though they were among the smallest of the original suite of plains predators, their strategy of hunting in packs could overcome an individual defense and made rapid retreat the most effective strategy. We'll never know for sure, but it's likely wolves killed more bison than humans hunting on foot and so were the primary predator that shaped *B. bison* behavior. In any case, since rapid retreat was also the best strategy against spear-throwing and, later, arrow-shooting humans, new antipredator tactics probably were important parts of the bison phylogenetic line's adaptation to new realities on the Great Plains.

MODERN RELATIVES

While bison in North America were moving from the steppe to the central grasslands, bison in Asia and Europe were withdrawing from the steppe to forest, becoming again adapted to forest edge and meadow. Today's descendant of that line is *B. bonasus,* the European bison.

You can tell at a glance that *B. bonasus* browses more than does *B. bison. B. bison*'s face is nearly perpendicular to the plane of the ground it walks on. It scarcely has to bend its neck to graze grass, but must strain a bit to lift its mouth to the right height and angle to clip leaves from a branch four or five feet above ground. The heads of European bison are set on their necks differently than those of their American relatives. The nose is further forward than the forehead when the neck is in neutral. It's easier for them to browse, but harder for them to graze. They're rangier than *B. bison,* and while their bodies have less hair, their tails have more. Their horns point forward through the plane of their faces as do the horns of all bison except *B. occidentalis* and *B. bison.* Forward-projecting horns facilitate fighting in the manner of domestic cattle—standing close and hooking. Straight-up horns facilitate the running charge, slam-heads-together fighting style of *B. bison.*

The two species are different enough that it's surprising to learn that they interbreed readily, producing perfectly fertile hybrids. In fact, the ease of interbreeding has created a conservation challenge. Some European bison were deliberately interbred with American bison to "refresh the blood" of tiny groups of European bison isolated in zoos. That genetic contamination has precluded using some of those zoo lines to increase the genetic diversity of the only free-ranging population, the several hundred bison in Poland's Bialowieza Forest.

In North America, *B. bison* may, or may not, have subdivided, splitting into a subspecies adapted to northern sedge meadow, *B. bison athabascae* (wood bison), and the more numerous group adapted to the Great Plains, *B. bison bison* (plains bison). Subspecies are a touchy taxonomic issue. By definition they are fully interfertile with one another; but if they are consistently different, especially qualitatively different (e.g., in color of feathers, shape of horns, distribution of hair), then underlying genetic differences are implied and subspecific status seems justified. Early observers described a different-looking bison living in the boreal forest of northern Canada and called them wood bison. A population of wood bison persisted through the near extinction of the plains bison. Wood Buffalo Park was created to protect them in about 5,000 square miles of boreal forest and sedge meadows well north of Edmonton. This population lived isolated in the northern part of Wood Buffalo Park until plains bison, their numbers grown beyond their pastures elsewhere in Canada, were introduced into the southern part of the park. The two populations mixed, and it appeared that there was no longer a pure line of wood bison. Then a small herd—just eighteen individuals—was discovered to the north. Some suspiciously well-worn trails connected the two areas, but the Canadian government finally concluded the herd was pure wood bison. It moved the herd further north, out of reach of the "hybrids," and nurtured the population, which has expanded well beyond a thousand individuals.

Even so, it's not clear that we have two subspecies in North America today. Supporters of subspecific status point to several ways in which wood and plains bison differ. Wood bison, they say, are taller and have more hair on their rumps and less on their shoulders and forelegs. Their hump descends smoothly both fore and aft from its highest point; in contrast, a plains bison's hump descends smoothly to the rear from its highest point, but dips and then rises again just a little as it descends forward to the neck. These taxonomists assert that the differences, though modest, are big and consistent enough. The differences are there. I've seen them—taken photos.

But there are some problems with this classification. The decision whether or not the differences between two sets of animals of one species are big enough to made them separate species is necessarily somewhat arbitrary, but everyone agrees that the differences must be genetically based. When geneticists look closely at the genes, using current technology to compare the blood types and the tiny sample of the DNA that are available, the genetic distance between woods and plains bison herds is small and not highly reliable. It is still possible there are substantial genetic differences not yet visible to the current genetic technology, but these results make that appear less likely.

The biologist Valerius Geist argues that the differences we see are deceiving. He says that rather than being genetic, they are the result of growing up in different environments. Thus, he says, the two "subspecies" are merely ecotypes. Many knowledgeable people don't agree. The issue is still open, and we're left with several possibilities: (1) Perhaps they were and are distinct enough to be subspecies. (2) Perhaps they once were but no longer are distinct. (After all, there were those suspiciously well-worn trails leading from the "hybrid" part of Wood Buffalo Park to the "isolated" wood bison population.) (3) Perhaps they never were distinct—at least not at the genetic level.

However the subspecies subplot is resolved, the major story line of the bison's last million years won't change. That is pretty well known and largely clear. And, until horses and hide hunters nearly ended it, it was an inspiring story. It's the story of a small, not particularly prepossessing member of the cow family that conquered the north's rigors, not only adapting to new challenges on a new continent but also becoming the dominant animal and a keystone species in North America's biggest biome. It's a Horatio Alger story for the ages. And the bison did all this while flouting Horace Greeley's advice: they went east.

CHAPTER 8 How Many?

The Bison Population in Primitive America

There was a long silence—a shocked silence, it turned out—on the other end of the line. "Thirty million? But Dr. Lott, we've always understood there were sixty million." A *National Geographic* researcher had called to hear my comments on a soon-to-be-published graph depicting the decline of the buffalo but was startled when I said the starting point of the graph was at least twice too high.

"My editor isn't going to want to take your word for *that*. Can you suggest some references we could check?" Sure, I said, call Jim Shaw. They did, and when the graph appeared, its starting point was 30 million. Publishing that number went beyond good research: it was an act of courage. Signs in parks and refuges, informational handouts, history books, newspapers—everywhere you look you get the same story: 60 million bison in primitive North America and as few as a thousand twenty years after the end of the Civil War.

Knowing how many there were goes far beyond casual curiosity. We can't understand the ecosystem of primitive North America, or the magnitude of the human rearrangement of that ecosystem, without a good estimate of primitive North America's bison population (see map 1). The available approaches are historical accounts and estimates of the maximum carrying capacity of the North American central grasslands. But it is terribly hard to get good numbers from either.

"Sixty million bison" has long been as close to religious dogma as a secular society's beliefs can be. At one time I presented it as gospel myself. Our faith in its reality is an important part of our view of our environmental and social history, and it has been used to quantify our ancestors' stewardship of the land they colonized. Such importance justifies a really close look at how that figure got fixed in our collective consciousness.

IN SEARCH OF THE 60 MILLION

Jim Shaw has traced the figure of 60 million back to the writings of Earnest Thomas Seton, the great naturalist, and to Seton's source—Colonel Richard I. Dodge.

Colonel Dodge, like his father and his father's father, was a career army officer. He'd been assigned to Kansas in 1869 and commanded Fort Dodge in 1872–73. In 1871 he took a trip along the Arkansas River in a light wagon. The total of 60 million has its foundation in how he described that trip.

Dodge said twenty-five miles of his thirty-four-mile trip passed through a herd traveling perpendicular to his route. Seton, or rather his four-volume *Lives of Game Animals,* was sitting on my bookshelves. The flyleaf of each volume is inscribed "Robert S. Norton"—my mother's father. I don't know when the book came into his hands, but it was published in 1927, four years before he became superintendent of the National Bison Range. I inherited it a quarter century ago. The 1927 work contained Seton's second effort to calculate the primitive America bison population. In a 1910 book he had started with an estimate of Dodge's herd that William Hornaday had published in 1887. Seventeen years later he started from scratch.

When I was a boy in Montana, a *figgerhead* was not a figurehead; it was someone with a head for figures. Seton was a figgerhead. Seton tried every reasonable way to estimate the number of bison in North America before the great slaughter. He read historical accounts of expeditions crossing the plains and prairies before the hide hunt. He determined as best he could the number of hides that had been shipped east by train during the slaughter; he tried to estimate how many bison the continent could feed. But in the end he fell back on the numbers he developed from Dodge's published account of his wagon ride.

Seton assumed a herd would travel twenty miles in a day, and as Dodge was in the herd most of the day he calculated that the herd Dodge traveled through was a rectangle twenty-five miles by at least twenty miles. He noted that buffalo could be as dense as twenty-five to the acre in large herds. I don't know how he discovered that, but to picture that

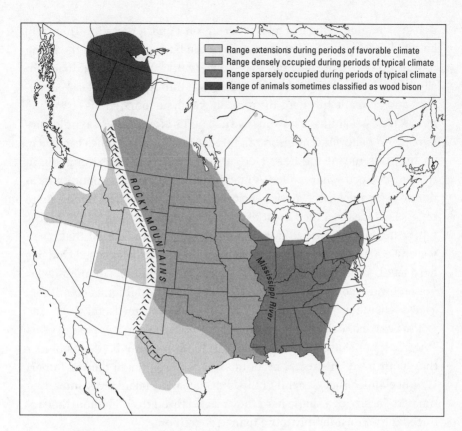

MAP 1. Plains bison distribution, about 1500

density, imagine living in a house that sits on a quarter-acre lot. There will be two buffalo in the front yard, two in the back yard, one in the living room, and one in the kitchen. To be conservative, Seton halved that density and calculated that the herd Dodge rode through was "at least" 4 million, and he made this herd the rock on which he founded his 60 million figure. He asked himself how many such herds North America could contain. Seton believed that bison migrated a few hundred miles each year. To him this meant the herd Dodge saw contained all the animals in roughly 200,000 square miles. A line drawn around every reported location of bison in North America enclosed an area of about 3 million square miles. This area included most of the Rocky Mountains and all of Idaho, where bison were rare to nonexistent. Seton acknowledged that bison weren't equally abundant everywhere, but he nevertheless used Dodge's herd as his model for all North America. Three million square miles is fifteen times 200,000 square miles. Fifteen times 4 million is 60 million. That's where the 60 million we've all heard about came from.

Jim Shaw has collected the various approaches taken over the 130 years since Colonel Dodge took his wagon ride. In the "my herd was bigger than your herd" historical account sweepstakes, a man named Robert Wright claimed that General Philip Sheridan and Major Henry Inman estimated the size of a single herd they passed through as a billion, then reduced it to "considerably more than" 100 million.

A question springs to mind. Was there enough grass in North America to feed 100 million bison? Or, as an ecologist would put it, "What was the carrying capacity of the whole area bison lived in?" In still other words, what's the biggest population the continent's bison habitat could have supported? That won't tell us how many were there—it just sets an upper limit; but given the scarcity of information, that would be a very useful thing to know.

All the efforts to answer the question in this form have limited themselves to the Great Plains—the central grassland that extends from the Rocky Mountains to the Mississippi River and from southern Alberta and Saskatchewan to central Texas. That's a reasonable approach, since by all accounts the great bulk of the bison population lived there. Estimates of the number that lived east of the Mississippi and west of the Rockies cluster around 2 to 4 million. There's no good reason to suppose there were more.

In 1910 the U.S. Department of Agriculture counted the livestock on the western grasslands. By 1910 forage production on the plains was enhanced by irrigation, fertilization, exotic plants like alfalfa, and agricultural strategies like putting up hay for winter feeding. And livestock have a lot of help getting through the winter—hay and grain to eat, structures to break the wind, containers of drinking water kept open when ice locks many natural bodies of water away. So even though livestock are less well adapted to North American winters than bison are, they can survive them better. All these agricultural arts were employed to raise carrying capacity, and they probably did so—boosting the possible maximum by an unknown amount.

There were 24 million horses and cattle and 6 million sheep on about half the Great Plains. Seton looked at these numbers too, did some extrapolating to other areas, and came up with a total of 75 million bison in primitive North America. To be conservative he lowered this estimate to a total of 50 million in North America at the beginning of the seventeenth century. But in the end he was so taken with Dodge's data that he went with that herd and held to 60 million. In 1991 Dan Flores reanalyzed the USDA's 1910 data county by county and generated a total North American bison carrying capacity of 28 to 30 million for a period of average rainfall. His base year fell in a period typical of the preceding 500 years. Drier times would have shriveled the grasses, the bison habitat, and thus too the bison population. For example, bison were either largely or totally absent from Texas and New Mexico during a dry period from 500 C.E. to 1300 C.E.

In 1972 the zoologist Tom McHugh used a different method to determine carrying capacity. He started with the total area of the central grasslands (1,250,000 square miles), then took "a range manager's" approach to keeping livestock numbers within carrying capacity. Using existing formulas developed for cattle, McHugh calculated conservatively to take dry years into account. He assumed that carrying capacity varied from one buffalo per 10 acres in the tallgrass prairie just west of the Mississippi to one per 45 acres in the short-grass prairie just east of the Rocky Mountains. He got an overall average of 25 acres per buffalo, or 26 buffalo per square mile. That gives the Great Plains a carrying capacity of 32 million bison. McHugh deducted 4 million for competing grazers—pronghorn, elk, and prairie dogs—and added 2 million for bison living elsewhere, for a final estimate of a maximum of 30 million buffalo on the continent in primitive times.

A more sophisticated variation on McHugh's approach has recently become possible. The Natural Resource Conservation Service has mapped the soils underlying the central grassland. It has gone further, providing formulas that estimate the year-long primary productivity of native grasses on particular soil in a particular place with a specific amount of precipitation.

Thomas Haynes combined this information to calculate the bison-carrying capacity of fourteen counties in southeastern Montana. It's a 35,000 square mile area—about the size of Indiana. Haynes assumed that the average bison weighed 1,200 pounds. (In a growing population, with lots of calves and yearlings, the average might well be a bit lower.) The carrying capacity was the number of 1,200-pound ruminants the plant community could feed given the forage that grasshoppers, nematodes, and other consumers would use up. Since rainfall varies considerably on the Great Plains, Haynes calculated this area's carrying capacity for wet years (1.4 million bison), normal years (1.1 million bison), and dry years (800,000 bison). Through the 1800s about one year in four was dry.

Though no one has done it, it would be possible to expand Haynes's approach to the whole of the Great Plains—at least that portion in the United States. It would seem that such a calculation would establish the minimum and maximum number of bison on the Great Plains. But it's not that simple. Suppose after one or more dry years Montana had a wet year. The population couldn't possibly increase from 800,000 to 1.4 million in one year. That growth would be 600,000 animals—a 75 percent increase. Even if all reproductive age cows had a calf, every calf lived, and no adults died, the population would grow about 40 percent, not the 75 percent it would take to get to the wet year carrying capacity. Population biologists call this a *recruitment lag*. The primary productivity of the short-grass prairie can drop by 90 percent in a dry year. Grasses would not recover from a bad year in one year and the cows coming off a bad year wouldn't have a 100 percent calf crop. Let's assume that realistically, somewhere between 50 and 75 percent of the cows would deliver calves.

Not all the calves would live; wolves would have eaten some of them. Ludwig Carbyn, Sebastian Oosenbrug, and Douglas Anions showed that when wolves are numerous relative to bison, they can take up to two-thirds of the calves. An educated guess is that they regularly took one-third. A third of a 50 percent calf crop is a third of a 20 percent population growth, or 7 percent of the population. This leaves 13 percent growth in a wet year if no adults die. Two-thirds of a 50 percent calf crop (20 percent population growth) takes 13 percent of the population, leaving a 7 percent population growth. One-third of a 75 percent calf crop leaves a 20 percent population growth if no adults die. Two-thirds of a 75 percent calf crop (a 30 percent population growth) takes 20 percent of the population, leaving 10 percent growth.

All the above analysis assumes that wolves are the only killers of calves and that no adults die. But adults do die. Wolves kill a few, mostly cows, during the winter, but winter weather itself can be a catastrophic killer. In 1841 a flow of warm air melted the top of the snow that lay on Wyoming's short-grass prairie. Then cold air froze the melted snow into a layer of ice that bison could not break through to get to the grass below. By spring there were millions of fresh bison bones in Wyoming, but no living bison. Those bones were never disturbed by bison hooves. If the population hadn't been sinking fast under the weight of hunting, dispersers from elsewhere eventually would have occupied the habitat. But as it was, buffalo never returned to Wyoming's grasslands. Mary Meagher, a Park Service biologist who has studied Yellowstone's bison for more than 40 years, has seen several winters kill 20 percent of the population.

A bison population can have much bigger busts than booms. In fact, no bison population has ever grown faster than 20 percent a year even when it had zero predation and negligible winter kills. Given the frequency of dry years and the reality of wolves, the likelihood that the bison population would ever have reached the wet year carrying capacity is vanishingly small. Suppose that the population eating partially recovered grasses and preyed on by wolves grew 10 percent a year. At the end of a dry year it would have reached a maximum of 800,000. In year two it would have increased to 880,000, and in year three 970,000. Through the 1800s about every fourth year was a dry year, so on the average the population would drop to 800,000 in the fourth year, never having risen above 88 percent of the grassland's carrying capacity for a normal year or above 69 percent of the carrying capacity for a wet year. Throw in an occasional catastrophe like that Wyoming winter kill, and the fact of recruitment lag probably kept the population well below the normal year carrying capacity.

That's more sophisticated speculation, but it's still speculation. We have no ways to count historical bison, and therefore can supply no answer that would satisfy a modern wildlife manager. About all we can confidently say is that primitive America's bison population was probably less than 30 million—perhaps, on average, 3 to 6 million less.

PART FOUR

The Bison's Neighborhood

All animals eat plants, directly or indirectly. Bison eat them directly, feeding on grasses. Grasses evolved relatively recently (for plants) and came to dominate the treeless savannas that began to emerge some 30 million years ago.

The great savanna between the Mississippi River and the Rocky Mountains—the Great Plains grassland—took its present form only after the last ice age ended around 10,000 years ago. During some periods a wetter climate supported trees and the area was partially wooded. But for most of the past 10,000 years it has been a grassland occupied by a peculiarly American suite of savanna inhabitants.

CHAPTER 9

The Central Grassland

Where Buffalo Roam When They're at Home

GRASSLAND ECOLOGY

From southern Alberta to central Texas, from the Rocky Mountains to the Mississippi River, a sea of grass covered the middle of North America—the area called the Great Plains covered 15 percent of the entire continent. The bulk of primitive America's bison population lived there. Plants conform to a simple general rule: where the clouds bring more water as rain and snow than the sun and wind can evaporate in an average year, trees grow. Grass grows where there is at least half as much precipitation as the sun and wind can evaporate. If there is so little precipitation or so much evaporation that grass can't grow, you are standing in a desert. In a temperate climate like that of the American Prairie, a very rough rule of thumb is that more than ten but less than forty inches of precipitation per year makes for a grassland. But while fourteen inches might be plenty in northern Montana, it might be too little in southern Texas.

The American Prairie is doubly rooted in the Rocky Mountains. The rise of the Rockies created a rain shadow favoring grassland over forest, and the soils the grasses grow in came largely from the Rockies. About half the original mass of today's Rocky Mountains has eroded. Water, often in the form of ice, did most of the eroding, and moving water carried many of the eroded particles east as far as the 100th meridian of longitude. The 100th meridian lies just east of Pierre, South Dakota, and hits Dodge City, Kansas, almost dead center—about halfway across the central plains. But there was another great conveyor belt at work too: moving air—the prairie winds. Loess is wind-borne silt. The prevailing westerlies carried loess to and beyond the Mississippi. Nebraska and Kansas were almost completely covered by a thin layer (less than five feet deep) of loess. Take away plant cover and roots, stir the loess, and it's ready to move again. What wind brings, the wind can take away. It was loess that the westerlies carried from the Midwest's dust bowl to the Atlantic in the 1930s.

The central grassland's native plants, along with dirt dwellers such as earthworms, modified those materials into fertile soil. But not always the same soil. Hudson Bay and the Gulf of Mexico are at sea level. The western edge of the American Prairie, just east of the foot of the Rocky Mountains, is nearly a mile high, and, located in the Rocky Mountains' rain shadow, pretty dry. Every mile downslope, to the east, it's lower, wetter, and warmer and the soil is darker—from brown in the west through chestnut to black at the eastern edge of the prairie.

Ecologists divide the American Prairie into three regions, each named for its predominant grass. Other things being equal, the closer to the Rockies a given location is, the less rain it receives and the shorter its grass. At the foot of the Rockies—say Denver, Colorado, or Billings, Montana—you're at the western edge of the short-grass prairie. Travel east to Kansas's western border, and you'll find the short-grass prairie blending with the mixed-grass prairie. It's a blurred and shifting border, not at all precise, but nevertheless important. The mixed-grass prairie attracted more bison than either of the other two. It covered all but the eastern edge of the north-south tier of states starting with North Dakota and extending through South Dakota, Nebraska, Kansas, and Oklahoma and into central Texas. The tallgrass prairie lies between the mixed-grass prairie and the Mississippi. During a long drought—say eight to ten years—the short-grass prairie moves east as its dry-adapted plants outcompete the taller, thirstier tallgrass prairie plants. When the rain returns, the mixed-grass boundary region shifts west again. Dividing the central grasslands into sections can make each seem small. It helps to remember that the short-grass prairie alone is about the size of western Europe.

The American Prairie's grasses are deeply and broadly rooted in the soil: established for the long run. They're perennials—their root systems are designed not for months but for decades. Only 10 percent of growth is above ground in leaves and seeds; 90 percent of growth is in the roots, and each plant sends about three-fourths of its carbon below ground into its roots. With so much energy stored below ground, the plant can persist through years-long droughts. In the wettest grassland, eight-foot-tall grasses such as Indian grass and big bluestem spread their hundreds of roots as far as six feet down into the deep black soil to tap the moisture accumulated there in past years. In the west, blue grama and buffalo grass send up stalks about ten inches tall and fill the soil below and beside them with roots outfitted with tiny hairs that absorb water from the upper thirty inches of soil, thereby capturing the moisture from even small storms. In between lies the mixed-grass prairie, a blend where the western short grasses grow in especially dry locations, the eastern tall grasses grow in especially wet locations, and intermediate, two-foot-tall grasses such as little bluestem and western wheat grass are widespread.

But there wasn't just an east-west, wet-dry gradient. Our grasslands also stretched thousands of miles north and south, from subarctic in southern Alberta to subtropical in central Texas. The length of the growing season—crucial to all plants—is as little as 100 days in the north and as much as 250 days in the south. Grasses adapted to the northern climate are called cool season grasses. In the south, warm season grasses dominate. A remarkably plastic species of grama grass stretches over the whole range. Local varieties of this species have adjusted their rate of development from seed. In identical conditions, seeds from the northernmost plants develop three times as fast as seeds from the southernmost.

Along the 100th meridian of longitude, the elevation is at 2,000 feet and the average annual rainfall has risen from sixteen inches in mile-high Denver to twenty inches in Pierre or Dodge City.

Plants respond to their immediate circumstances, not their longitude. So a focus on longitude alone suggests that things are simpler than they really are. The Staked Plains of Texas, the southernmost reach of the central grasslands, is *flat*. Thousands of square miles of flatness. But the rest of the central grasslands roll gently to steeply, creating dry, well-drained soil on the tops of low hills and wetlands or streams in the swales or ravines between them. There's less evaporation on north-facing slopes than on south-facing. The south side of a particular hill in eastern Colorado might support a short-grass prairie community while the north side supported a mixed-grass community and tall grasses grew in the wet bottoms at the foot of the north slope. Rivers such as the Platte and the Arkansas meandered east from the foot of the Rockies in wide riverbed bottoms nourishing water-loving trees such as cottonwood and cedar, as well as water-loving grasses from the tallgrass prairie. These wet bottoms were only about 7 percent of the total plains habitat, but they were critical for wintering bison and, later, for wintering plains Indian bison hunters.

Water, soil, and grass: bison were never far from the basics. They walked on the soil that nourished their nourishment. The water that made grass grow soaked their hair as rain, or accumulated on it as snow. Often the rain was accompanied by lightning, the sire of fire that forced the grass to start again from the roots up, sending forth tender, green, nourishing leaves that drew bison like a magnet. Thus come together the three great drivers of grassland ecology—precipitation, fire, and grazing.

All three great drivers were swayed by the wind. On the Great Plains they say that trees lean east and people lean west. Wind tilts every aspect of Great Plains weather. Spinning winds become tornadoes. Vertical winds pile up thunderheads. These negatively charged clouds may connect to earth via fire-starting lightning, and may release either refreshing rain—or hail. In the winter winds snowflakes fall sideways for miles, making the world ten feet away invisible at high noon in a blizzard—a word that was invented to describe winter storms on the Great Plains. Cattle, descended from wild stock in warmer, calmer climes, turn their backs on this onslaught and drift until a fence, or another barrier that accumulates both cattle and snow, stops them. There they remain imprisoned until either they or the winds die.

An individual buffalo is affected little by these kinds of winds. It can dodge tornadoes. Thunder and lightning don't scare it, hail doesn't hurt it much, and it faces into a blizzard equipped with the dense mass of curly hair on head and forequarters and the patience inherited from generations before it. Yet to bison as a species, wind made a big difference. Warm chinook winds from over the mountains to the west created the first open pastures as winter waned. But the most important wind for bison was one they never felt: the polar jet stream, a wavy rope of rushing air circling the northern temperate zone four to six miles above the Earth, along the southern edge of the mass of cool, dry air from the Arctic. The jet stream marks the northern limit to warm, moist air pushed north from the Gulf of Mexico by Bermuda highs, and it sometimes creates low-pressure, storm-inducing areas. When warm, wet air from the Gulf of Mexico saturated the plains, grass grew abundantly, bison ate heartily, and the populations rose. But sometimes for a year or so the jet stream was well to the south, bringing dry air over the plains. Then as now, every five to ten years eastern Montana would have endured fifty or more consecutive days in the growing season without rain. The grass would have retrenched. Only 20 percent of a short-grass prairie plot in western Kansas was bare soil in 1932. After four years of drought, 97 percent of it was bare soil. The drought dealt a body blow to bison. Cows went a year or more without conceiving, and more of the now leaner bison died each winter.

It seems hard. For each buffalo the shift in weather meant hunger, less chance of reproducing, more chance of dying. For the bison as a whole it meant a shrinking population. But the dry years were what kept the prairie a grassland. If every year were wet, trees would grow, grass would go, and the bison would follow. That's not to say that those who suffered in the droughts did so for the greater good of bisonhood; they were just unlucky. But their bad luck was an inescapable part of the boom-and-bust cycle of all temperate grasslands. And it's the bust part of that cycle that made sure the minerals from their bones nourished a grassland covered with living bison and not a woodland haunted by their ghosts.

The wind-rippled grassland whose surface undulates from horizon to horizon strongly evokes a sea, but it's a sea that can catch fire. Grass burns all in an instant. A dry stem glows red and turns to curling ash while you are still drawing a breath. When a wind pushes it, a prairie fire runs fast. The American Prairie has always burned. For millions of years lightning caused combustion, but people began to burn the prairie several thousand years ago—often with buffalo in mind. Sometimes they used the flames to herd the bison, driving them to a place for easier and safer killing by people on foot. Sometimes they burned the grass so that new growth would attract bison to a more convenient killing place. Growth-stimulating fires were set in the spring when new growth would quickly replace the old. In the fall, new growth was up to six months away. Bison deserted the bare ground created by a fall fire until spring, and the people who hunted that ground faced starvation.

Grass burns only when the aboveground part of the plant is fairly dry. Fire then consumes everything, leaving bare ground with a light coat of ashes. In tallgrass prairie, where mature grasses and their litter intercept 99 percent of the sun's rays before they reach the ground, fire creates a moment when a short plant's leaves can nourish their roots. The plant community that rises from the ashes is richer in species and more complex. The grasses surge back from their roots—the soil two inches down would not have been warmed even two degrees Fahrenheit by the fire—but the new leaves are different from those whose ashes they rise through. Enough nitrogen is carried away in the smoke so that the new growth has a higher ratio of carbon to nitrogen. That makes it poorer forage than was the now-burned grass when it was newly growing, but better forage than the mature plants that burned. Fire decreases production in the short-grass prairies, where water limits growth. But in the tall grasses, where the failure of radiation to reach the soil limits growth, fire increases production.

Fire interacts with grazing. In the mixed-grass prairie, little bluestem presents grazers with an in-your-face defense—stiff tillers (stalks) that the grazer must push through to get to the green leaves. Fire removes the tillers, and bison, which avoid tillered little bluestem, graze the new-growing little bluestem as readily as other grasses. The fire that consumed little bluestem's defenses thus helped little bluestem's competitors through the mechanism of bison grazing. The grasslands are as much creatures of the grazers as the grazers are creatures of the grasslands.

Other, smaller, grazers also played a role. Grasshoppers, for example, were always present, and occasionally a tidal wave of Rocky Mountain locusts would roll east with the westerly wind, eating every blade of grass in a swath a hundred miles wide and hundreds of miles long. And below the surface nematodes, nibbling at the roots, ate more grass than everything else put together. Still, in their heyday, bison were big on the plains—big enough to be called a keystone species.

Grassland has a reciprocal relationship with bison, though the reciprocity is somewhat roundabout. Bison aren't very picky, but given the choice they will choose grass over forbs (i.e., herbs other than grass). That small preference makes a big difference to the grassland. In the tallgrass prairie, grazing bison keep the dominant tall grasses such as big bluestem and Indian grass short enough so other species can also grow. Consequently there are more plants representing more species where bison graze.

Grasses resist being eaten. The tall grasses outgrow the grazers. In a few weeks they become tough, unpalatable, and protein poor. In the West, where there isn't enough water to outgrow the grazers, buffalo grass employs the opposite strategy. It grows too short to be grazed easily, keeping its leaves low and tucking them back and down where they're hard to reach. Grazing costs the grazed grasses much of their leaf structure—the photosynthetic, energy-producing part of the plant—and the plants react. They boost the photosynthesis of the remaining and replacement leaves to compensate. In the short term this strategy makes up for the lost tissue. In the long term there's no free photosynthesis: the grasses boost their short-term output by dipping into their capital of stored nitrogen, and as that gets drawn down they're less and less able to compensate. They need about two years' rest to recharge their carbohydrate and nitrogen batteries from a big draw-down.

Bison affect species composition in two ways. First, they wander. When they had the whole prairie to wander over, particular patches of grass probably had two-year rests fairly regularly, especially as bison choose areas where grasses are growing most vigorously. When the tallgrass canopy is grazed off, the sunlight reaches the earth and the shorter plants do better. There are fewer individuals of more species after grazing—just as after fire. But grazing-stimulated growth in the western short grasses tends to eclipse smaller plants. Grazed short-grass prairie has more individuals of fewer species.

Second, bison don't just take away. They give something back—fertilizer. From a prairie plant's point of view, urine is a bath of nitrogen dissolved in water—the answer to its prayers. The grasses' leaves and stems would have eventually decomposed and returned the nitrogen to the soil, but after a longer delay and in a form that the plant would spend more energy using. Bison drawn to the close-clipped, nitrogen-rich grass in colonies of black-tailed prairie dogs leave a disproportionate amount of their digestive by-products there, thus transferring nitrogen from the rest of the prairie soil to prairie dog towns.

Bison don't just graze and eliminate on a prairie, they also wallow and die there. The soil from the wallow probably gets rid of some insects, possibly reduces the bison's heat load, and certainly changes 75 to 150 square feet of habitat for the prairie plants. Wallowing lays the soil bare and compacts it. The compacted bowl of soil holds rainwater, creating a microenvironment in which seeds can sprout and seedlings of plants—sedges and rushes in tallgrass prairie—that are otherwise rare on the prairie can grow. Some of these seeds are blown in by the prairie winds, others are carried there in the coats of the wallowing bison—perhaps picked up in another wallow. Bison maintain old wallows for years. Ecologists have even found wallow-shaped and -sized depressions in prairie soil 125 years after the last bison left a locality. Pleasure in the discovery of one of these "relict wallows" is a bit more subdued these days, for careful analysis of the underlying soil shows that at least some are depressions caused by very local geological processes perfectly independent of buffalo. Too bad.

But if there is ambiguity about where bison wallowed a long time ago, there is none about where one has died recently. Any flesh that isn't consumed or carried away decays. A whole carcass is soon leaking fluids that poison the prairie plants, denuding the leaked-into area. But though it is toxic at first, the decayed bison eventually enriches the soil. First the flesh and then the bones yield minerals and organic matter, creating a patch with two to three times the nitrogen in the surrounding soil up to three years later. The first plants to invade the few square feet of habitat are soon followed by the perennial grasses, but in the meantime otherwise rare species have a place to go to seed.

GRASSLAND PAST AND PRESENT

When Lewis and Clark headed west from the eastern edge of the central grassland, they were exploring not a wilderness but a vast pasture managed by and for Native Americans. Domestic horses had come to the grasslands a century earlier. They had put the whole region in hunters' reach, and tribes that had hunted buffalo along river bottoms or forest edges now rode the length and breadth of the grassland. They used fire as a management tool, burning small areas in the spring to attract bison. Their management must have changed the details of the landscape but did not fundamentally alter its nature, the plants that grew there, or the wildlife that lived there.

We humans are creatures of grassland—taking our first bipedal steps across grasslands when they were, geologically speaking, nearly new themselves. Might not the habitat that gave birth to our species ring some dimly remembered bell and reassure us—lift our spirits? The Serengeti is a grassland and one of the most spirit-lifting places on earth. George Schaller, who studied lions there for several years, described it as "a region of light and space." Beautiful and true, but not the whole truth. It's also a teeming kaleidoscope of lions, antelope, elephants, cheetahs, giraffes, zebras, hyenas, and dozens more species. Herbivores and their predators are part of a grassland as fish are part of a sea. With them it is vital, rich, reassuring.

For all that, the American grasslands with their teeming bison herds and wolf packs did not reassure European-originated, tree-minded Americans as they traveled west. Early explorers called the western half of this fabulously productive grassland the Great American Desert. These travelers were preoccupied not with what was there but with what was missing. They couldn't see the grassland for the lack of trees. And because they couldn't see that what was there was productive, they set out to replace it. A few tried farming the western plains in the 1860s. They were cautious, especially west of the 100th meridian in the "Great American Desert."

The plow brought night to the grassland's day. The whole central grassland defied the first plows put to it. It had been there for millennia, and its roots were abundant and strong. When the strength of those roots was tested by cast-iron plows, the plows often broke first. But in 1837 John Deere invented the steel plow. Steel plows broke the roots, with a sound like a pistol shot. Those pistol shots were music to the plowman's ears—the score to a sort of anthem of grassland settlement. The lyrics of this siren song were "Rain follows the plow." Standing meteorology and ecology on their head, this song said the sparse rainfall on the plains was not the cause of the grassland but an effect of it. If the grassland were plowed and trees and crops planted, then rainfall would increase dramatically. William Gilpin—army officer, western traveler as early as 1843, a man of boundless optimism for the future of the Great Plains—was the best-known singer of this song, but he had an impressive backup chorus. Frontier folklore had long voiced the same view, and some government geologists chimed in. With this song in their hearts, pioneer cultivators fell upon the land like starving men upon food. Nature seemed to have prepared the way for them—for the first time in the move west there were no trees to fell, no stumps to clear. A modern ecologist would have taken the absence of trees as an ominous sign, but even if they had known and agreed with the sign, the plowmen would have been undeterred. Of course there was too little rain for trees—the land had not yet been plowed. Plow it and the rains would come.

So they plowed it, and the rains did follow. Through the 1870s the jet stream must have flowed well north. Wet year followed wet year just as farmers put plows to the plains. How farmers and their families must have celebrated Gilpin and the folklore he propounded on their first dozen Fourths of July on the plains—rain had followed the plow. In the wheat states—Nebraska, Kansas, the Dakotas, and Minnesota—the population soared from less than a million in 1870 to more than 2.5 million in 1880. Wheat farmers were replacing buffalo. But though the plow had preceded the rain, it had not caused it; and when the jet stream moved back south in 1886 it heralded a ten-year drought. That Fourth of July was celebrated west of the 100th meridian among stunted and prematurely brown fields of grain. Hope and faith shriveled with the crops. Homesteads were abandoned, suicides were buried, and the land began to be gathered in ever-larger plots in ever-fewer hands—a process that still continues.

Tallgrass prairie was largely converted to corn, another tall grass that Native Americans had cultivated there, by 1900. Too little rain fell in the mixed-grass and short-grass prairies to raise corn. Wheat was the domestic grass that came closest to being adapted to these regions. The Great Plow Up began in the West to feed Europe during World War I, but it accelerated after the war as America became a grain-exporting country. Wheat acreage in Montana grew from 250,000 acres in 1909 to 3.5 million acres in 1919. From 1914 to 1919 Kansas, Colorado, Nebraska, Oklahoma, and Texas increased their wheat acreage by 13.5 million acres. The plow up continued until 1935, when the rains failed, the wind didn't, and the dust bowl was born.

More than 99 percent of the tallgrass prairie has been plowed; most of the world's corn and much of its wheat grow there. The more arid the land, the less of it has been plowed. About 42 percent of mixed-grass prairie and about 29 percent of short-grass prairie have been converted either to cropland or to nonnative grass pastures.

The native perennial grasslands made soil; an annual grassland spends it. A clump of native perennial bunchgrass eighteen inches across may have two miles of roots. They cling to the soil and it clings to them. That's not true of many of the domesticated grasses that have been planted in its place—wheat, oats, barley, and corn are annual plants. They have evolved a radically different strategy. They live only long enough to produce a single crop of seeds. Instead of storing energy in their roots, they put it into their seeds. Since those are the parts of grasses we humans generally use most, we prefer annuals. But their roots are minimal, die each year, and don't hold much soil. Raising cereal crops on the Great Plains trades soil for seeds. Today's annual grasslands are spending the capital that the native grasslands had banked. Grain growers have ways to slow the erosion—crop rotation; cultivation that follows the hill's contours; the alternation of strips of crop with strips of fallow, moisture-accumulating soil plowed at ninety degrees to the wind—but nothing yet stops it, let alone reverses it. Despite all the arts of modern agriculture, on most of the plains west of the 100th meridian a money profit is possible only by running a soil deficit.

It's still possible for a few people to make a living most years, and to feed many more from the dwindling soil. But someday—someday soon in much of Great Plains country—there will be too little soil to produce food profitably. What then? Some people have a vision in which sizable parts of it go back to feeding buffalo, and they want to do it before so much soil is gone that the land will be barren.

A VISION OF THE FUTURE

The starting point for all these visions is a problem: what we're doing in the area west of the 100th meridian these days isn't working very well. The visionaries have proposed a very conservative solution—go back to what worked in the past. The most efficient way to hold the remaining soil and to build more is to re-create the perennial grasslands. But while the native grasses will hold and build soil, they won't yield human food. They'll yield buffalo food, though, and people can eat buffalo. In *Bring Back the Buffalo!* (1996), Ernest Callenbach recommends that basic strategy—let the prairie grasses grow again, let the bison eat them again, let people eat the bison again. And, he adds, let the western prairie winds turn turbines and generate electricity instead of blowing topsoil toward the Atlantic.

Native grasses won't return until some political and economic arrangement is worked out that will support the change. Before the plains were settled, the region bison roamed was called the *buffalo common*. Commons in England and colonial America were pastures where every member of the community had the right to graze their own livestock. On the buffalo common, nobody owned the livestock while they were alive. Individuals became owners of the animals by killing them. A new buffalo common has emerged in the minds of two geographers and urban planners.

Frank and Deborah Popper noticed that large portions of the Great Plains were losing people. They suggested replacing what they saw as failing farming with a buffalo common. As academicians sometimes will, they became quite specific about which counties in which states seemed ripe for conversion to buffalo common status. That got people's attention, and it made a lot of them mad. NIMBY is not just an urban phenomenon. The Poppers must have blanched at the uproar, but they didn't shrink from it. Instead they made their case over and over, right where it was least welcome—in the states and counties at issue.

The ensuing debate about the future of the semiarid parts of the West was a proper dust up, but it has also cleared the air in many ways. It got people looking at such things as the magnitude and significance of population loss, and the realities of time and money needed to restore a native grassland in the face of lost soil and of introduced plants and plant pests. There are more buffalo on the Great Plains today than when the Poppers started—all the increase is in domestic stock on private land. That's well and good. A domestic line of bison would be gentler on the short-grass prairie than either wheat or cattle. They'll walk further for food and trample stream banks less; even the more cup-shaped bottoms of their hooves shift the soil they step on a bit instead of just compressing it. Through them the grassland can produce food without being plowed, and thus without being washed away. Perhaps America's mammal has a future on America's Great Plains as the twenty-first century's contribution to the stock of domestic mammals.

I'd applaud that, but I'd applaud it without forgetting that this development won't do anything for the bison as a wild animal. If it isn't done right it could harm wild bison. We must insulate wild bison by isolating them from domesticated bison. One place to do it right is in a Great Plains grassland park. If we have the will we can easily create a place where the Great Plains grassland nourishes and is nourished by the Great Plains' community of animals, with wild bison fulfilling their keystone role.

PART FIVE

The Bison's Neighbors

*Bison biomass dominated North America's central grasslands,
but they were not alone out there. The central grasslands supported a rich community of vertebrates, many of them influencing and influenced by the bison. A few of the other species — elk,
wolves, grizzly bears, and golden eagles — inhabited Asia,
Europe, or both. But most existed only in North America,
though not all — not even the bison — were limited to the
central grasslands.*

*As little as 150 years ago, when America's prairie grasslands
states and provinces were safari destinations for wealthy Europeans, the assemblage was still intact. Buffalo Bill could
show it all to the Grand Duke Alexis. But since then, the
community has been enormously reduced in size and its parts
somewhat scattered. Wolves and elk are gone entirely; wild
bison persist in a few tiny islands surrounded by fences
rather than water.*

*Today the only place we can see the whole community
assembled, and observe its interactions, is in our
imagination. To help that imagining I have assembled part
of that community on the following pages. I've had to be
very selective — a reasonable introduction to the whole community would require several volumes the size of this one.
Quite arbitrarily I've stayed close to the earth, ignoring a
fascinating array of birds except for one closely associated*

with buffalo and one that nests underground. Yet even this small sample allows a peek into the rich and fascinating ways of life that evolved in the heart of America and throve mightily there a century and a half ago. It gives us some small sense of what we are missing today and may encourage us to think of a way to get it back.

CHAPTER 10   Wolves and Bison

Myths and Realities

It was late spring in Wood Buffalo Park, Northwest Territories, Canada. Fifty to sixty bison, many of them mothers with calves, were cautiously approaching a water hole when the wolves rushed them. The bison turned tail and ran.

For a quarter of a mile they fled along a narrow trail through dense willows, and the wolves followed closely behind the last set of hooves, unable to move up beside the bison. But then the protective wall of willows ended at a meadow. Part of the herd ran close beside the trees on their right. As they swung right, three cows and calves became separated on the left, and the wolves pounced. This was a big pack, more than twenty, and in minutes the three calves were down and dead. A few days later the same wolves ran the same gambit at the same place and killed two more calves. Wolves don't always get their calf. Ludwig Carbyn, Sebastian Oosenbrug, and Douglas Anions, who watched these attacks, have seen calves that were down and covered with wolves bounce up like a buckskin-colored ball and escape. But it doesn't happen often.

All of us nature lovers are nuts about wolves. When they were reintroduced to Yellowstone Park they immediately became kings of the beasts there—a sighting of them is a great prize won. They have drawn a set of wolf watchers who spend hundreds of hours along roadsides each year, trying for just a glimpse. I joined the watchers for a few hours recently. I didn't get a glimpse, and I have never seen a wolf in the wild. I heard wild wolves howl at Hook Lake in Canada's Northwest Territories one summer, and watched Jack Smith examine dozens of their scats there. Ninety percent of those scats were full of young calves' rusty orange hair. We couldn't be sure the wolves had killed the calves. They might have found them dead and scavenged them. But it looked suspiciously like the low calf survival was due to wolf predation—like one of those situations where a predator takes so many young from a small population of prey that increase is impossible and population survival becomes doubtful. Wolves can do that.

Where wolves and bison coexist, wolves often closely follow a herd. Sometimes, though, the wolves are just snatching up small mammals that the bison's feet have flushed. Wolves are opportunists, able to capture and eat meat from mouse to moose. But while wolves will eat any morsel of meat they encounter, they're designed, body and behavior, to kill hoofed animals, even those as large as bison. Like the bison they follow and sometimes feed on, gray wolves survived the Pleistocene extinction some 10,000 years ago in North America. Before that they shared the North American plains with a suite of large predators. There was the long-legged, fast-running short-faced bear, the American lion, and the American cheetah—a 400-pounder that was probably as fast as today's African cheetah. There were several species of saber-toothed cats and a gray wolf relative, the more robust dire wolf.

Not that the gray wolf is a wimp. The wolves that hunt bison in Canada weigh well over a hundred pounds. Still, it was among the smallest of that suite of plains predators, yet was the largest to persist, as modern bison—smaller than their ancestors and pre-Pleistocene contemporaries such as mammoths and mastodons—were the largest hoofed grazers to survive. In my mind's eye I see these two species walking silently together through a gray fog of extinction with larger predators and grazers dissolving into the mist around them. Apparently they were just small enough to get through, so that each is now the largest survivor of the suite it belonged to.

Cats have tremendous jaws and kill large prey by suffocation or by biting through their skulls. None of the dog family has the mouth power to kill large animals quickly. So they chew their large prey to death, more or less eating them alive. It's very hard for one dog to do that, and nearly all the many species of dogs are social and hunt in packs—almost always permanent family groups.

Ah, family! With all its idealized connotations of safety, nurturing, tenderness, altruism, loyalty, and love, the very word warms us and so warms our feelings toward the wolf. Wild dogs, especially the big wild dogs, are famously family oriented, and wolves are no exception. Hunting parties are made up of families collaborating in the hunt, sharing in the kill, and, if the pack has pups at home, carrying food to them in their stomachs. Family is valuable and so is valued. Among dogs, the family that preys together stays together. Still, wolf social life little resembles the American dream family. Idealized American family life is about happiness. Wolf family life is about survival and reproduction. In most circumstances, each family stakes out and defends dozens of square miles to hunt in. Trespassing wolves will be challenged, pursued, perhaps killed.

There *are* affectionate greetings, nudgings, grooming among pack members; they even share food. But life within the pack is intensely, even relentlessly, structured, and all attempts to deviate from the established order are punished severely. Relationships are adjusted through physical, sometimes deadly, force. It puzzles me a bit that we nature-loving Americans, most of whom treasure political equality, have such affection for an animal whose social organization is basically a cruel despotism. There's no equality among wolves. One member of each dyad will be the tyrant and the other will be the tyrannized.

At the top stands the alpha wolf, tyrant of all, tyrannized by none. A step below is the beta, tyrannizing all but the alpha and tyrannized only by him. At the bottom is a wolf that tyrannizes none and is tyrannized by all. It eats last, if at all, and is casually bullied by all other pack members many times each day. For those at the bottom of the hierarchy it's a dog's life in the worst sense. But that's my perspective—the perspective of a Jacksonian democrat, focused on the underwolf. Viewing wolves from it is, in an important sense, simply silly. Judging wolves by our standards is as foolish as judging ourselves by theirs. The lone wolf is a romantic figure but a biological dead end, so even underwolves stay with the pack. They have no choice. For wild wolves, loyalty is life.

Political equality was not a factor in the evolution of wolf society. Producing and raising young wolves was. Even in a large pack there is usually only one father and one mother. They monopolize the reproductive opportunities—to use one of the economic analogies that behavioral ecologists find useful. But Patti Moehlman has pointed out that in dog species, the larger the litter, the more attending adults the mother needs to rear them. The other members of the pack, who are nearly always the dominant pair's siblings or offspring, are the attending adults. For the subordinate, membership in the pack holds out the possibility of promotion to a top spot. Meanwhile aunts and uncles or brothers and sisters bring food back to the den and baby-sit the growing pups. Caring for pups born to their parents or siblings increases the representation of the family genetic complement in the next generation, so they're also having a bit of reproductive success, though diluted by the distance of their kinship.

For many millennia some of the food that wolf aunts, uncles, and older siblings regurgitated to their young relatives came from young buffalo. We humans love baby everything, even skunks, so we may be a little put off by wolves picking on defenseless calves. But that's just the point. The calves *are* defenseless, and few other bison ever are. Wolves can kills cows, and do in the winter. They can even kill a bull in winter. But cows and bulls can injure or kill wolves. If a wolf is to make its living killing bison, it must choose a bison to kill every few days for the rest of its life. Those who choose best live longest and reproduce most. Wolves aren't wicked, they're just adapted.

Mature bison bulls are 2,000-pound athletes with sharp horns, hooves like sledgehammers, and a very short fuse. Their hide is thick, tough, inedible, and hard to chew through to get to the animal's edible parts. On the other hand, calves are small and scared, with button horns and hooves so soft that a wolf can chew them up and swallow them, along with the hair, hide, stomach full of milk, and all the bones except the jaws and the skull cap. There isn't much to eat on a calf, but what there is is choice. In Wood Buffalo Park, where the wolves live largely on bison, they kill only calves right through the summer. By fall half or more of each spring's calves have fed wolves.

It's not likely that the wolves are doing this population some good by weeding out the unfit, unless being young and unlucky is a form of unfitness. The calves being weeded out are the unlucky plus, perhaps, a few that are a little short of specifically wolf-resistant traits—for example, always keeping somebody else between you and the wolves, or always having a tough and resourceful adult at your side. When mixed herds—cows and bulls together—travel, cows and their calves tend to journey in the safest area, front and center.

Many of the survivors owe their lives to an alert and aggressive mother. She may have had help with her wolf problem, though probably not from another cow. When the wolves come it's usually every cow for her own calf. But occasionally she can get help from bulls. During a marathon standoff in Wood Buffalo Park, four wolves spent eleven hours trying to kill the one calf in a small herd—the calf's mother and fifteen bulls. Time and again the cow led the calf to, or the calf on its own ran to, one or more bulls. The bulls then charged the wolves, and sometimes surrounded and accompanied the calf. In the end, although the wolves were able to get their teeth on the calf eleven times, it was very much alive—even frisky—the next day.

Every calf has a father. Is it he that gallops to the rescue when the wolves are at his child's throat? The buffalo bull that stands between his calf and all the wolves in the world is a sweet myth. I really like this myth: it makes the bull the kind of father we all wanted, and some of us had. For one thing, "My old man can whip your old man" would be a claim hard to dispute. But more than that, the demonstration of selfless commitment to his calf makes life seem safer and surer. Like other myths it expresses our longings, and so reveals us to ourselves.

But while it helps us understand ourselves, it doesn't help us understand bison. The father is almost certainly somewhere else, alone or with a few other absentee fathers—enjoying the warm sunshine, eating the green grass, resting, ruminating, eyeing the competition, and generally getting ready to defeat it. Selection has focused him. His spring work is to get ready for the summer rut. Anything that distracted him from that task, even defending a calf against wolves, would be a dereliction of destiny. The competition is fierce during the rut, and a small edge has a big payoff. During the eleven-hour siege in Wood Buffalo Park it was young bulls—too young to breed in the coming rut, let alone to have bred in last year's rut—that most kept the wolves at bay.

The bull of our fond myth, the attentive, protective father who spent the spring defending a calf or calves, would arrive at the rut underweight and tired. There he would compete with another kind of bull, the inattentive father, who would arrive fat, fast, and fierce. If an attentive bull wanted a calf to look after next spring he would have to adopt one sired by an inattentive bull, because he would be unlikely to sire one himself that summer. Thus inattentive bull bison have more offspring, and their sons behave like their fathers.

Many real animals have a mythical one that stands beside them in our imagination. The real animal is a product of natural selection; the mythical one is a product of our yearnings and fears. The real one can teach us about nature; the mythical one can teach us about human nature. As long as we can stay clear about the difference, each can teach us a lot. But if we confuse myth with reality, the more we learn the less we will understand about either the world or ourselves. The two bison have much to teach us about both.

with horses,
add domestication

CHAPTER 11   Buffalo Birds

It's May in the Wichita Mountains Refuge. The wind jostles my pickup gently, and sundown is coming soon. Twenty feet before me on a green flush of spring grass I watch two animals. One is a three-year-old bison bull. The other is a black, shiny male buffalo bird (we call them cowbirds nowadays), about the size of a blackbird. The bison is grazing—a step, several bites, another step.

Really rich grouse hunters in Scotland employ beaters: people who march in a line, thrashing the bushes with sticks to flush the grouse from hiding and give the hunters a shot at them. But these birds were using buffalo as beaters long before there were shotguns or even gunpowder. The bird walks along a few inches from the bison's muzzle. There are hundreds of animals I can't see—insects in the grass under the bison's feet and mouth. The grazing bison flushes them and gives the buffalo bird a shot at eating them.

It's spring and all over the Great Plains birds are feathering their nests, filling them with eggs, or feeding hatched babies. But the male I'm watching will eat the insects the bison flushes rather than feeding them to its babies, and the females accompanying the feet and muzzles of nearby bison will do the same. As their eggs become ready to lay, female buffalo birds scout around for a place to lay them in already feathered nests containing newly laid eggs—always of another species, because buffalo birds don't make nests. So they lay their eggs in the nests of strangers who, if luck holds, will incubate their eggs and feed their babies. The bird people call such birds *brood parasites*, and they show up in the best of families. English cuckoos always display this behavior. Black ducks always do it. Goldeneye ducks sometimes do it. It's an intriguing way of life, but it has its hazards. Some of the nest owners recognize the bad egg and toss it out. But for the buffalo birds the approach gets around a big problem. Bison can travel tens of miles overnight; for the birds, that changes a five-minute trip to the corner grocery to a long search for a new food source that may be hours away. Buffalo bird nestlings need to eat most of their weight every day. The logistics of depending on beater buffalo for food makes rearing your brood yourself a chancy business. So traveling with the herd and being a brood parasite go together nicely.

Which came first, the beater buffalo or the parasitically laid egg? We may never have an answer to that, but the way of life raises some other questions that we can answer. For example, how do you know who you are if you never see a member of your species while you're growing up? Most goslings and ducklings that see only people from the egg on follow people when little, and may court them when grown. A buffalo bird leaves the nest without ever having seen one of its own kind, but it knows a buffalo bird when it sees one and always mates with one.

And then there's the song they sing. We humans tend to sing the songs we heard in our youth the rest of our lives—if not the exact songs, then very similar ones. Most singing birds we know about are like us, singing songs much like the songs they heard in the nest. But buffalo birds sing a buffalo bird song, not the song the male who fed them sang, and not a song they are likely ever to have heard. Has selection protected them from foreign influence by making them impervious to experience? Well, not quite. While he sings a perfectly recognizable buffalo bird song, tutoring can improve it; and he gets his tutoring not from the male who feeds him but from his prospective mates and his bachelor buddies.

A buffalo bird that has never heard the song can sing it. But there are many ways to sing any song, and he is open to suggestions. In fact, he is looking for them. He perches a foot or so from a female, fixes an intense gaze on her, and sings his song, first this way, then that, trying out styles. The song is short. At its end the female usually sits unmoved, and he sings again in a different style. But once in a while she flips a wing, ever so slightly, at the end of the song. The male is mildly electrified. He quickly repeats the song she just flipped over; and when she flips a few more times after hearing it, that becomes the song he sings.

With his song style perfected, he seems well launched on the royal road to romance, or at least the freeway to fertilization. But the song may get vetted one more time—by his bachelor buddies. He sings his song around them too. It may be a multipurpose song, saying he's a fighter as well as a lover. His buddies' reaction depends on his status. If he is the dominant male in the group, they just listen. But if he's not, those that dominate him attack, and they keep on attacking after each song until he sings a less effective song—one that females are less likely to flip over.

This second vetting biases the breeding toward the dominant males, because the songs a female flips to are the songs she will stand for when the singer tries to mount her and fertilize her eggs as they form. The DNA he sends and she accepts, like the DNA it pairs with, contains the music and lyrics, as well as the means to change the style of singing it—for better and for worse.

Somewhere else in all that DNA are instructions for finding bison and using them as beaters. The instructions aren't so detailed that domestic cattle won't do now that the bison are gone. But though they've now been cowbirds for dozens of generations, and most domestic cattle don't go anywhere overnight, not a single buffalo bird has ever opted for the burdens and the opportunities of home, hearth, and family.

CHAPTER 12

Diseases and Parasites

A bison's big body is the largest repository native to North America of the sun's energy converted to flesh and blood. The size and numbers of their bodies made bison too big a resource to ignore, and many creatures have found a way to get a bit of the sun's energy by tapping into the bison's store. True, their sheer bigness thwarts many—coyotes, say, and even wolves—that might try to harvest this stored energy. But predators aren't the only exploiters of bison. Like all big organisms they are a resource for hundreds of kinds of tiny life forms that use them in a variety of intriguing ways. And size is no protection at all from very small things—it just makes you a bigger target.

Liking bison as I do, I tend to look at diseases and parasites from the bison's perspective: that is, as an affliction. But then, from a grassland's point of view, bison are, at least at times, an affliction. These tiny life forms inhabit bison as bison inhabit the prairie. The adaptations by which they manage themselves and their resource are in some ways as fascinating as the bison's adaptations. An introduction to the whole cast would fill volumes, but you can't really know bison without meeting a few of the actors using them as their stage.

MICROBES

The smallest of the organisms acting on the bison are single-celled creatures so tiny that their relationship to a bison is like the bison's to an isolated landmass—an island or a continent—that has a suitable habitat but is a long and dangerous way from home. When these microbes colonize a bison it's a bit like bison dispersing from Asia to North America. Most incipient colonizations fail. Just as the immune system resists incoming

microbes, so the initial animal invaders are usually destroyed or out-
competed by animals already present. But if the new animal or microbe
manages to become established, it may reproduce explosively as it either
encounters no competition in its niche, overwhelms the competition it encounters, or, in the case of disease organisms, evades or overwhelms the
host's immune system.

Both kinds of invaders disperse from the initial colony, and generation
by generation they occupy the available habitat. Most of the microbes that
have occupied a habitable part of a bison settle into some sort of long-
term relationship with it. A bison can't digest grass without the help of
bacteria in its gut. Disease organisms don't do it any good, but few actually destroy it. After all, the bison they are occupying is both their home
and their progeny's launching pad to the next bison. For while they may
occupy any one individual for thousands of generations, it will eventually die—in effect sinking like the legendary Atlantis. The microbe line
will die with its host if the microbe doesn't use either that bison's life or
its death to send descendants on to the next host.

Most, like *Brucella abortus,* use their host's life to send their descendants
on to the next bison. The genus *Brucella* gets its name from its discoverer,
Sir David Bruce. In 1887 Bruce isolated *Brucella melitensis,* the cause of
Mediterranean fever in humans. Ten years later Bernhard Bang identified
Brucella abortus in cattle. It's often called "Bang's disease" or simply
"bangs," and an infected herd is sometimes called a "banger herd."

Brucella abortus takes its species name from one of the ways it uses its
host's life to get its young into the next generation: it sometimes causes
abortion. The infection is called *brucellosis*, and the fact that some wild
bison have it is at the swirling center of disputes about how wild bison
should be managed—indeed disputes about whether or not there should
be any wild bison. *Brucella abortus* came to North America from Europe,
inhabiting domestic cattle, and first began to inhabit bison in Yellowstone
Park around 1917. It now infects many individuals of the two largest herds
of wild bison remaining—those in Yellowstone Park in the United States
and those in Wood Buffalo Park in Canada.

Brucellosis in the Yellowstone bison population has stimulated a controversy as hot as the rocks that heat Old Faithful's water. Like that water,

the controversy regularly boils over and erupts. It's too hot to touch comfortably, but too salient to ignore. Therefore I will briefly sketch the controversy before describing the fascinating and successful way of life of the single-celled creature at its core. My sketch of the controversy won't end with *the* solution to this problem, or even *a* solution. Either requires expertise I don't have—expertise in microbiology, immunology, and epidemiology. But what we do regarding these herds won't be determined by biological expertise alone. At bottom, wildlife management in our society uses biological knowledge to implement individual values as they are expressed through our political system. I *am* an expert on my own values, and I don't hesitate to advocate them.

Every winter hundreds, sometimes many hundreds, of bison are shot on Yellowstone's western boundary. This policy is designed to prevent their exposing to brucellosis the cattle that graze every summer on the largely public lands that surround Yellowstone. Those lands, together with the park itself, form an ecological unit called the Greater Yellowstone Area—commonly abbreviated as GYA.

Many bison in Yellowstone Park have brucellosis, and it is possible that those bison could transmit the disease to cattle if they occupied the same range at the same time. A number of state and federal veterinarians are pressing to completely eradicate brucellosis from the United States. As brucellosis currently occurs only in the GYA, the GYA is the focus of their campaign. Some have taken the position that only total eradication is acceptable, and they threaten to penalize the cattle industry in Montana and Wyoming if brucellosis persists in the GYA.

Brucellosis can infect a wide range of mammals, some birds, and even some insects, but it tends to die out in most species. It's transmitted from one species to another when an animal eats forage contaminated by an aborted fetus or contacts the fetus itself. The fluids in a colonized cow's aborted fetus are loaded with *B. abortus*—billions of microbes in a teaspoonful. Any mammal that gets these fluids in its mouth, nose, or even eyes can be infected. (Humans are a secondary host. We develop a fluctuating fever called "undulant fever"—my father's mother got it as a child by drinking unpasturized milk.) A curious young cow, bison or domestic, that licks the fetus or eats plants the fetus has contaminated can easily ingest enough *B. abortus* to found a new colony.

Brucella abortus is designed to survive the digestive tract and enter the blood stream. An infected animal's immune system attacks the intruder and sometimes eliminates it. An aborted fetus can be infectious for a few days in summer, or a few weeks in winter, though in the wild that much high-quality protein is unlikely to go uneaten by scavengers for more than a day or two. Brucellosis is not a catastrophic disease. Except for the occasional abortion, its symptoms are mild to nonexistent in infected cows, and it poses no meaningful threat to humans today.

Cattle graze on much of the GYA that is in public ownership outside the park. Bison are currently allowed to graze only inside the park. Elk graze in both areas and move between them freely and regularly. Cattle, elk, and bison all have the potential to transmit brucellosis to one another, and elk-to-cattle transmission has been demonstrated in a few cases.

Broadly speaking, the two sides of this controversy are both concerned with managing the risk of transmission to cattle. One approach is to eliminate it completely by eliminating brucellosis in the GYA, using whatever means are necessary. The other is to lower the risk to the extent possible without doing damage to wildlife values in the GYA.

Brucellosis has been eliminated from cattle by testing every animal, destroying those that test positive, and vaccinating calves—usually just the female calves. Even though no test is 100 percent accurate and no vaccine is 100 percent effective, this strategy will eventually eliminate brucellosis from a confined herd. Some have proposed using this approach to brucellosis in Yellowstone's bison. Management would corral every bison in Yellowstone, restrain it to draw blood, give it a unique mark—probably a numbered ear tag—and keep it corralled until the test results were known. Individuals that couldn't be corralled would be shot.

There are problems with this method. Corralling and handling the animals would be inherently traumatic. In addition, the capture-and-handle regimen would have to last several years and would select against wildness in the Yellowstone bison for generations. There's also a problem with using tests and vaccines developed for cattle on bison. Both tests and vaccines are most effective on cattle, and both are less than 100 percent effective even on cattle. Their success rate is lower in bison, and we don't know how much lower. Finally, there is history. The Yellowstone herd went through exactly this sequence more than half a century ago but was soon reinfected—undoubtedly by elk.

Though the out-migration of infected elk doesn't draw the fire, either figurative or literal, that bison out-migration does, brucellosis can't be eradicated in Yellowstone's bison without eradicating it in elk. If the test-and-slaughter approach that has been used in cattle were applied to both bison and elk, it would remove enough herbivores from the GYA to profoundly alter the ecological community for decades to come. Some believe, and can cite evidence in support of their belief, that brucellosis will not persist in elk living as wild animals without contact with infected bison or cattle. But that evidence is not strong enough to justify a program that depended on it. Moreover, such a program would require keeping a purged herd of bison off all elk habitat—essentially living in corrals—for generations.

The eradication approach gives absolute priority to eliminating risk of cattle infection. From this perspective, the only acceptable risk level is zero. With current technology, that can be achieved only at a high cost to the population and genetic characteristics of GYA wildlife.

The alternative is to make the risk to cattle as small as possible while also minimizing cost to wildlife and the GYA ecosystem. Foremost among the available strategies are (1) excluding bison and elk from all areas where cattle ever range; (2) excluding cattle from public lands in the GYA; (3) letting the three species overlap, as cattle and elk do now, thus exposing cattle to brucellosis but protecting them against infection by vaccinating those that graze in the GYA; and (4) letting the three species overlap, but grazing only steers on the GYA. Steers could catch brucellosis, but they could not transmit it. All four are feasible containment strategies. Both options three and four would allow infected wildlife and cattle to overlap in the GYA. Both add a modest cost to grazing in the GYA, but they eliminate an enormous cost to its wildlife on what are, almost entirely, public lands.

The future may bring technology that will enable us to eliminate brucellosis in the GYA without damaging its wildlife. There's a vigorous search for better technology, especially for better vaccines and less intrusive ways of delivering them—for example, via a dart gun or oral administration in food. These hold out the hope of at least lowering the infection rate in bison and elk without the genetic and ecological costs imposed by the eradication currently being contemplated.

As I said a few pages back, wildlife management uses biological expertise to implement the personal values of individual members of our society as they are expressed through the political system. Just now, management must tilt, at least a little, either toward cattle in the GYA or toward bison. I vote for a tilt toward bison. That vote isn't based on superior understanding of the issues. It's a statement about where my heart lies.

Now let's return to brucellosis itself, changing focus from the controversy that swirls around it to the adaptations that permit this single cell to expand its unwelcome domain despite our sophisticated resistance and the even more sophisticated resistance of its hosts' immune systems.

While an aborted fetus is the *B. abortus* strategy that worries cattlemen, it's not the usual way in which *B. abortus* seeks a new habitat. *B. abortus* can survive in an aborted fetus only a few days in spring or a few weeks in winter. This method of propagation is a lot like launching a raft of colonists into unknown ocean currents with a week's food and hoping they make a favorable landfall before they starve. A better bet is to wait for the fetus to develop and then establish a colony in the calf. Establishing a colony in the calf is a pretty sure thing, because the whole *Brucella* clan specializes on mammals—animals that nurse their young. No surprise then that *B. abortus* thrives in the mammary glands, where it enters the milk and then rides the milk run from the mother's udder to the calf's stomach. With the cow's udder pushing and the calf's mouth pulling, the forces are truly with the *B. abortus* colonists.

It's a great colonization system, easier by far than the bison walk across Beringia, the land bridge that emerges between Siberia and Alaska during an ice age. And while *B. abortus* must wait more generations (about 100,000, as active *B. abortus* have offspring about every twenty minutes) for a calf to become a mother than bison had to wait for a new ice age to reveal a bridge, every infected cow's calf will be colonized and most will not be able to clear the infection. It spreads not quite as fast as a chain letter, but with a good deal more certainty. Yet it's not absolutely certain. The cow's immune system fights back, attacking this foreign body. *B. abortus* has counterstrategies, including hiding out in the cow's white blood cells. Even so, the immune system may eradicate *B. abortus* altogether or suppress it so effectively there are no longer enough cells to colonize a new individual. Still, daughters that survive to breed will create more new habitat and, if their immune systems haven't eradicated *B. abortus*, ensure its colonization.

The longer a cow lives and the healthier she is, the more calves (and occasional aborted fetuses) she is likely to produce. *B. abortus*'s invariable toast should be "To our hostess. May she live a long and fruitful life," and in *B. abortus*'s list of rules for behavior toward the cow, rule number one should be "First, do no harm." *B. abortus* does no harm. So far as we know, it doesn't affect the cow's health at all. There probably have been *B. abortus* strains that injured their hosts, but such injury would have reduced transmission—a losing strategy—and they have lost out to the benign *B. abortus* we see today.

Indeed, most diseases go easy on their host, but anthrax, which arrived in North America from Europe around 1800, is a killer. *Bacillus anthracis* is the bacterium and it kills to colonize. It travels from habitable host to habitable host in a capsule built by the single cell that the capsule encloses. This capsule can sustain its life in a hostile environment (the only friendly environment is the inside of an animal) for as long as forty years. And we think space suits are a big deal.

It seeks its host as a spore, lying encapsulated in the soil. Summer rain has flooded it, winter cold has frozen it, bison may have trampled it, and still it waits. The search can last for what would be hundreds of thousands of generations of active cells. To colonize the lungs, its usual route, anthrax must rise, changing from a soil contaminant to an air pollutant. Usually its moment comes when the soil turns to dust and the dust is lifted into the air—maybe by the wind blowing, maybe by a bison pawing the ground or wallowing. A bison breathes in the dust and all in a moment a new island is colonized, the bacillus is released from its capsule, and the launching of millions more capsules is only two or three days away.

The way *B. anthracis* uses a bison's life to colonize another host is dramatic and daring. It destroys the bison in a kind of slow explosion, and uses the energy to propel its spores into a search for a new set of lungs. It's as though some race of humanoids from a sci-fi author's imagination invaded a livable planet while encased in a sophisticated space suit of its own manufacture (maybe in an arrested metabolic state), drawn passively to the planet by gravity. Emerging from the space suit it reproduces rapidly, plants an enormous atomic bomb in the heart of the planet, makes space suits for everybody, then sets off the bomb. The bomb propels its clan and the planet's debris into space, where each space-suited humanoid may drift for a million years until either another planet's gravity draws it in or it dies. Of course, *B. anthracis* doesn't blow its bison to bits— it kills it with a lethal combination of poisonous proteins (exotoxins), and it kills it fast. An adult bison will die two or three days after colonization. As the bison decomposes, insects and other scavengers consume this enormous store of energy, freeing the again encapsulated *B. anthracis*— multiplied millions of times—to the soil from which it emerged. Dust to dust in about two hundred generations.

TICKS

Ticks quest. *Questing* is a rich word, bringing to my mind either a passionate pilgrim or a noble knight crisscrossing medieval Europe seeking goodness and justice. But it's also how parasitologists describe a hungry tick scrabbling across soil and up plants—seeking blood. Blood to ticks is like grass to bison. It supports all their aspirations. It's food for growth, and—when they are full grown and ready to make babies—food for courtship and for males to make sperm. Females use blood to form eggs and to store within those eggs enough energy for their young to grow large and strong—able to latch on to a passing blood supply and start growing. Questing at first seems an odd term to describe a creature seeking another's blood, but I like it. It shows that the parasitologists are looking at the process from the tick's point of view.

The kind of blood sought depends on the kind of tick—they tend to specialize a bit. The one most likely to go for bison blood is *Dermacentor albipictus,* the winter tick. The winter tick quests for most of North America's large mammals, including bison. Much larger and longer lived than bacteria, much smaller and shorter lived than wolves, ticks use bison as a place to grow up. From the tick's point of view, a buffalo is an effectively infinite bucket of blood from which, once you manage to insert your straw, you can suck a lifetime's supply of nourishment. But first you have to catch your buffalo. It's not easy being a tick, and few quests are successful. It's mostly a waiting game.

Take a larva's first step: find a place where a buffalo will pass within reach of your front legs. On a prairie, the tip of a tall stem of grass is a good bet. Climb to the tip and hang on until either your minute supply of body moisture dries out, turning you into a tiny cornflake, or your prey brushes against your perch and you grab on. A tick must be a discriminating grabber-on. If it grabs just anything within reach, it could end up clinging to a tumbleweed and headed for biological oblivion. But ticks aren't clueless. When something comes within reach, they check it for signs of life. They respond to warmth, sounds, carbon dioxide, and even pheromones released by already resident ticks that are in a mating mood. If it's alive and they can reach it, they cling to it.

From questing to clinging is a big step toward reproducing, but there are still other steps ahead. First, the ticks must manage to hang on despite not being at all welcome. They're serious parasites, after all—a single tick growing to maturity on a farmer's calf costs the calf a pound and a half in growth. So the tick's targets resist. A buffalo's first line of defense is the hair coat that stands between the tick and its skin, and even for a tiny larva—the form winter ticks arrive in—this is a formidable barrier. Bison are very hairy. They have more primary hairs per square inch than any other members of their family—ten times as many as cattle—and a woolly undercoat as well. So even with a firm grip on the bison's hair, the tick is on the outside trying to get in. Time is short because bison do many things that tend to terminate the relationship. The bison's second line of defense is good grooming. They wallow, covering the tick with suffocating dust or scraping it off on the soil. They rub against trees or bushes, challenging the tick's grip. They scratch their neck and head, even reaching between their horns, with their hind feet. And, probably the most formidable defense of all, the tongue or teeth sweep across the hair on paths dozens of ticks wide. It's a wonder that any ticks break through to safety. In fact, few do. Bison have a small fraction of the ticks suffered by elk or moose living in the same habitat.

Desperate as things are for a tick that grabs a mature bull or cow, they are much worse for the tick that happens to hitch its hopes to a calf. The biologist Mike Mooring spent more than a year quantifying how much time each age and sex class of bison spent grooming during each season. Calves groom up to sixteen times as much as adults. It's not because they have the energy to spare, it's because they *don't* have the energy to spare. A tick takes the same amount of energy from any bison, but that's a much larger proportion of a calf's total energy. The smaller the animal, whether it belongs to a small species or is simply young, the greater the relative cost per tick and the more vigorously the host employs anti-tick strategies. Hence, among bison, calves are the most inhospitable hosts.

But bison can't spend all their time defending themselves from ticks. They have to budget their time and energy to cover a lot of essential activities. How do they know how much time and energy to spend attacking ticks in each season? It seems intuitively obvious that they feel the ticks in their hair or on their skin, and scratch more when they itch more, but in fact that's not so. The winter tick, unlike most North American ticks, hatches and attaches in the fall. In October the mature ticks have left to lay eggs and the minute larvae are wandering about on and in the hair coat: few have reached the skin. This is when bison itch least, yet it's when they scratch most. Their grooming steps up sharply in the fall when it will be most effective but is least stimulated by what's happening on the skin. So their grooming rate is endogenous—governed from within. It is something that long-ago bison fighting long-ago ticks somehow encoded and passed on. Bison calves aren't alone in employing this grooming clock trick against ticks. The biologists Ben and Lynette Hart (Ben was Mike's mentor) had already demonstrated it in several African antelopes. And Mike found that moose and elk also groom when that grooming is least stimulated but most effective.

Endogenous sounds like "in the genes" but it may not be. Not only each species as it evolves, but each individual as it develops constantly interacts with and is modified by its environment. Ben Hart suggests that a species-specific grooming clock is genetically programmed, but the grooming rate is modulated by such things as hormones, tick saliva, and so on, by which it adapts to local conditions. So while a six-month-old buffalo calf that grooms furiously its first October isn't grooming at a rate set by ticks on its skin, the grooming rate may still be calibrated to the local abundance of ticks. Perhaps its brain has a grooming-rate setting that adjusts to the amount of tick antigens the mother's immune system produced during pregnancy or lactation. That would allow the calf to fine-tune its budgets of time, energy, and attention to tick numbers. It sounds a bit complicated, but behavior is subtle and animals often surprise us with the ways they assess and adapt to their challenges. Whenever an animal passes up the chance to adjust its effort to the work to be done, it's failing to optimize. Certainly there's an upper limit to the fine-tuning they can do, but starting out with the assumption they might be able to make a particular adjustment is a research strategy that has revealed sometimes astonishingly rich behavioral capacities.

That's as true of parasites as of their prey. The winter tick leads a simple life—at least compared to other ticks. It's a one-host tick: a tick that attaches to a bison as a larva stays there while it molts first to a nymph, then to an adult, and leaves only as a corpse or a fertilized female looking for a site to lay several thousand eggs. If the tick gets past the buffalo's defenses to its skin, it makes a small, painless hole in the skin, fixes its mouth on the opening, and secretes a glue that cements its head to the skin around that opening. Then it "engorges," slowly filling its distensible stomach with blood. But even the winter tick's simple life has its complexities. Those ticks that evade the bison's defenses can grow up alone, but when it comes time to mate they need to meet somebody. Males detach and go in search of females, which are still feeding. But the females aren't passive. They've refined their come-hither strategy far beyond Hollywood's imaginings. Hollywood hints that mating is all about what's seen and heard. Winter ticks use the old reliable—smell. A maker of perfumes who could produce scents as powerful and action-specific as a female tick's scents could guarantee a man would be truly enslaved by the scents of a woman, and would be richer than Bill Gates.

When a mature female starts to engorge she also starts emitting an attractant sex pheromone. Release it and they will come. A male's odor receptors are located on his front legs, and he waves them about to orient himself to the source of the siren smell. But when he gets to the female he hesitates. This female could be early in the feeding cycle, not yet close to laying eggs, or she could even belong to another species. The parcel of sperm he has prepared is precious. He must deposit it where it will count. If the female is ripe she emits another pheromone—the mounting pheromone—and the male, as completely controlled by external forces as a dandelion seed blown in the wind, mounts her. Yet still he hesitates, holding on to his sperm until the female releases the last pheromone, the one that only their species makes—the genital pheromone. Now he places his store of sperm in her genital opening. After she uses them to fertilize several thousand eggs, she detaches herself, drops off the bison, and lays the eggs in a sheltered place—under leaves, in a crack or crevice. There they will hatch into larvae, and begin *their* quest.

Buffalo are like the prairie they evolved to thrive on—habitat to a whole community of living things, influencing and influenced by each of them. Some make the animals healthy: bison could not digest grass efficiently without the colonies of bacteria in their rumens and intestines. Others make them sick, even kill them. Some, like winter ticks, are ancient inhabitants; others, like brucellosis and anthrax, are newcomers; but all are part of the bison's world and their economy, as the bison are part of the prairie's.

CHAPTER 13

Pronghorn

When the bison were breeding on the National Bison Range, they usually absorbed all my attention. But in the quieter hours another sound sometimes registered in my mind between the roars and snorts of the rutting bulls—a wheezy nasal snort that seemed to echo itself as it was repeated in descending pitch. One particular day the pronghorn buck vocalist was "Graybuck," a male in the prime of life named for the gray tint of his white markings. He was standing on a ridge, and as usual, he was the only adult male there. But he was with a group of females and their fawns. His head was up and his far-seeing eyes swept over several square miles of grassland. Other large males, "master bucks," were dispersed over the grassland, each hundreds of yards from his nearest neighbor. Groups of females and their young drifted along, each group accompanied by an eagerly, even frantically, attentive male. Feeding as they went, the females Graybuck accompanied drifted north and the nearest neighboring male in that direction positioned himself in their path.

As the does came closer to the neighbor, Graybuck moved round in front and tried to herd them away. But they dodged around him, as they usually did, and soon reached the neighbor. Graybuck and the neighbor were suddenly face to face as the females drifted on past. Graybuck started to follow, but the neighbor blocked his path. Graybuck headed west and the neighbor walked parallel to him, ten feet away. They stopped and stared at each other, turned east and walked in parallel again, each on his own side of an invisible boundary. After a few minutes Graybuck turned away and walked to the south. The neighbor hurried to catch up with the females, who were moving northeast through his territory, toward where another male waited a few hundred yards away.

Graybuck, now alone, drifted along, stopping to forage occasionally. When he came to a patch of bare ground a couple of feet long and half as wide he sniffed it, pawed it with one forefoot, straddled it, and urinated in it. Then he stepped forward to defecate in the same spot, showering down a handful of pellets that came to rest among hundreds of others he'd already dropped there.

Those pellets were all his and, in a certain sense, that place was his. True enough, a few coyotes or badgers and bison by the hundreds may pass through it or camp out there, but no pronghorn could be there except at his sufferance. It was his territory. He patrolled its boundaries, chased away intruders, and put his smell on it. In fact, to a more sensitive nose than ours, the place must have fairly reeked of him. Besides his urine and feces, he made it smell of him by anointing some of the plants with musky oil from a gland in his cheek. He'd take a tall stalk of mullein or goatweed into his mouth and then, having wet it, apply his scent from his cheek gland. Several times I've rushed right over to a just-scented plant and sniffed eagerly. I can't smell anything, but many humans can; and if an olfactorily challenged species like ours can catch a scent, it's a safe bet that another pronghorn inhales a wealth of information from both the marked plant and the pawed dirt: who left the scent, how long ago, maybe even something about his physiological state—and the warning that any other males will be challenged here. The males had started all this in spring, and they would keep it up until September. Then, during a frantic two weeks, each mature female will take an interest in sex for a few hours—for the male attending the group just then, a moment worth seizing.

Those moments are what all the trespassing hullabaloo was about—life lavished on the possibility of making more of the same kind of life as completely, though not as abruptly, as a Pacific salmon swimming home to spawn for the first time and dying a few days later. But why spend life this way? Bison never defend any space. "Stay away from me and my cow and you won't get hurt" applies equally to all bison everywhere. And why make a big to-do about a particular area? Grassland's all the same, isn't it? Well, it is and it isn't. A bison eats grass, and a grassland is more or less equally grassy everywhere, so to a bison every part of it is pretty much the same. A grazing bison favors grass but is not picky.

But a pronghorn *is* picky—and has to be. Small bodies need less food, but they also need better food: more protein, fewer carbos, less lignin. Generally, the smaller a warm-blooded animal is, the more of its body's warmth is lost to the air and, to compensate, the higher its basal metabolism must be. There is such a thing as a better class of grass. The growing part has more protein. So younger is better and the part nearest the roots is better, but what makes life possible for the pronghorn is a supply of forbs—small broadleaf plants growing among the grasses. They have more protein and less lignin. The pronghorn picks them out, one or two at a time, from the surrounding grass. From a field that's 95 percent grass and 5 percent forbs, a pronghorn will eat half grass and half forbs. Bison have a wide mouth, so a mouthful may include several plants. Pronghorn have a narrow mouth that selects plants one at a time.

The pronghorns' perspective on forbs makes their relationship to a grassland very different from the bison's. For the bison it's just grass going on forever, but for the pronghorn there are patches of forbs growing among the grasses—more forbs in lower-lying, wetter ground and fewer on higher, drier, sunnier ground. The amount of food in such a patch is small, but so is a pronghorn. These patches make some parts of the grassland more attractive to pronghorn than other parts. Females spend more time in the better patches; so the better the patches the male defends, the better his chances of being the only male in the right place at the right time.

The ecologist Dave Kitchen followed the pronghorn on the National Bison Range for several years. He measured plants to see which territory had the most high-quality forage and he counted the number of does each buck bred with. Overall, the richer the territory, the more does. Animals in general are more likely to be territorial when resources are patchy. And that's so whether the patches are food or females. Often the two are connected—females spend their time where the food is, and males spend their time where the females are. That's as true for bison as it is for pronghorn; the difference for the bulls and bucks is that cows and does eat different diets, and the cows' food is not very patchy.

One midsummer day several years later, I watched a big buck on the National Bison Range going through the marking motions—sniffing, pawing, urinating, defecating, cheek-patch marking—but they appeared to be irrelevant. Groups of does wandered everywhere unaccompanied by an owner and not anticipated by a buck at a boundary. I was seeing the last remnant of territoriality; it dwindled all summer and died away completely before breeding began.

The ecologist John Byers, who studies behavioral development, realized that the territorial pronghorn Dave had described offered an opportunity to watch young males develop into territorial master bucks. He got there just a little too late. The territorial system was fading away. John was already a seasoned field-worker and found a way to turn disaster into opportunity—he documented the disappearance of territoriality while he and Dave collaborated in a search for the cause.

Breeding in September is hard on the bucks. They lose a lot of weight, some are injured, and there is little time to recover before winter sets in. Some of the master bucks die every winter: the worse the winter, the more that die. The winter of 1978–79 was horrendous, and all the bucks that had held territories the summer before—the master bucks—died. Moreover, the survivors were all young—juveniles, really.

In the following years those youngsters grew up, and there soon were as many old males on the range as there had been when most of the grassland was defended and does bred on territories. But while the population had recovered, the territorial tradition hadn't. Perhaps they were too young or too similar in age, for in 1982, the first year in which John collected data after the master buck die-off, only a fourth of the territories were held through the season, and by 1984 none were. Yet some thread ran through whatever passes for a pronghorn buck's mind. In 1984 six bucks established a territory, though none maintained it through the season. That number is way down from the fifteen that established and maintained a territory the summer before the big winter die-off, but something was still stirring.

This is a real brain bender. Should we say social organization is really just a product of culture? I don't think so, but it's clearly not just a product of ecology and of age and sex classes. We must acknowledge the power of culture in some form. Part of the story may be cultural inertia—manifested here as the cost of being out of step with the rest of society. It's a lot more cost-effective to be territorial when everybody is being territorial. A territory becomes available when the owner dies, moves on, or is evicted. You and all or most of the other big bucks have a space and stay in it. You mark it fairly often and occasionally defend it against the tentative encroachment of a neighbor, who usually retreats peacefully. Once in a while a band of bachelors passes through, but not one of them is a match for you and they are easily hurried on their way. You're busy, but not frantically so, and it's mostly just a matter of showing the flag and staying home. When the females drift in, you have them to yourself and it's all been worth doing.

But if there are no territories already in place, trying to hold one is a different deal. A lot of guys your size are drifting about and their intrusions are not tentative. By the time one is evicted, two more may have arrived. If you spend so much time defending your forbs that there is too little time left to eat them, you will go into a downward spiral, becoming weaker while the intruders get stronger and harder to evict. Often they're still there when the females pass through. Little wonder that you eventually pull the plug on your territorial strategy. The greater wonder is that you try it again next spring. Something must be echoing in your head from the generations of pronghorn before you that sets your feet on that path. It's as though there was an urge, but not an irresistible one. An urge that may dwindle as things get tougher or as you get weaker—of such fluctuating urges, social plasticity must somehow be made.

Pronghorn territoriality on the National Bison Range seems to have died a demographic death in 1978–79. But in most species in most situations, the abundance and distribution of food are crucial determinants of territoriality. The behaviorist Chris Maher decided to find out if that was also true of pronghorn. Her first step was to examine the literature. In eleven of the twenty-three published reports, pronghorn were said to be territorial; in the other twelve they were reported as not territorial.

But wait. Suppose that territoriality, like beauty, lies mainly in the eye of the beholder? If it does, might all twenty-three reports be describing the same thing from different perspectives or tastes? Standards for what is beautiful will always be subjective—as much a reflection of the beholder as the beheld. But standards for what is territoriality shouldn't be. To be objective, science needs beholder-independent standards. One of the rocks on which science founds its objectivity is numbers, or, put more precisely, quantification. Kitchen and Byers furnished some numbers—how many bucks started territories in what years and how long they held them—and since the same people compared one time to another and saw different things at different times, we can be confident that there was a difference. But how can we be confident that different observers, each of whom saw only one thing, and reported different things, weren't seeing the same thing through different eyes? If one friend says it's warm where she is and the other says it's cool where he is, you don't really know whether the temperature is higher where she is or she has different standards for "warm" and "cool." But it's easy to resolve—find out what a thermometer reads in each place at that moment. To compare variation in social systems we need behavioral thermometers.

Chris developed a thermometer-like scale to measure pronghorn social systems. The scoring was a sort of formalization by quantifying the things bucks do to mark place and defend space. Chris counted the number of times each male approached an intruder per hour, the percentage of time he spent cheek-marking, and his number of linked urination-defecations, snort-wheezes, and chases per hour. These numbers enabled her to relate territorial behavior numbers to some others—the number of inches of rainfall and numbers describing the plant community. She quantified species diversity, species richness, percentage of forb coverage, dry weight (an index of plant biomass), and percentage of leaf nitrogen in two places—eastern Montana and Wind Cave National Park, South Dakota. The more rainfall, the more plants; and the more plants (especially forbs) in summer, the higher the territorial temperature—that is, the more the bucks cheek-marked, snort-wheezed, challenged intruders, and so forth. We can see some biological logic in these adjustments. When resources are spread thin, defending them could cost more energy than a buck can afford.

And Chris was able to give us some insight into the physiology of pronghorn social system change. She collected males' droppings at both places and measured their testosterone. Where males were more territorial, their testosterone concentrations were higher. It's a bit of a tease though. A correlation doesn't show what causes what, just that two things occur together. In other species it works both ways—in many birds, the more testosterone a territorial male has the quicker he is to challenge an intruder, and challenging intruders also increases his testosterone. It might work either way or both ways in pronghorn.

I sometimes wonder if there are any circumstances in which bison would become territorial. Big grazers that simply eat everything that grows on the ground rarely do, but there are intriguing exceptions. The white rhino, which was named for its mowing machine mouth (*weid* means "wide" in Afrikaans but was corrupted into "white" by English speakers), often lives in an elaborate territorial system. Even horses, which don't ruminate and are adapted to a very low-quality diet, became territorial off the Carolina coast. Stallions defended each end of a narrow island, creating a territory several times longer than it was wide and approachable only via a short boundary whose length was the middle of the island's width. Territory defense was so cheap that territoriality could be made to pay.

I can't really imagine bison defending territories, but my imagination has been smaller than their repertoire before.

CHAPTER 14

Prairie Dogs

"Did the earth move for you too, my prairie dog princess?"
"Of course it moved, stupid! There's a buffalo wallowing on the front mound."

Black-tailed prairie dogs are small as individuals; each weighs only about a pound and a half. But their numbers—perhaps 5 billion little more than a century ago—made them loom large in the prairie's economy. That abundance of flesh, blood, and burrows attracted a host of predators, parasites, and hangers-on. Prairie dogs were in many ways as central to the prairie economy as bison, but unlike bison they lived not just on the prairie but in it as well. Prairies have no trees, and thus prairie dogs spent their lives literally under the feet of bison. And it's not just that both live on the prairie and so are bound to be in the same place at the same time once in a while. Prairie dogs live in "towns"; and when you build a prairie dog town, bison will come.

Actually, a prairie dog town is more dug than built. They create tunnels too small for most predators to enter and so make homes that are more secure and also, being underground, more temperate. They use the excavated soil to build a wind-powered, flow-through ventilation system. Their burrow usually has a back door anyway, in case of hostile intruders. When the wind blows, its speed at the top of the mound is faster than at the ground-level entrance because the friction of ground and grass slows the wind so much that it's significantly faster just a few inches above the surface.

Prairie dogs shape the soil excavated from the burrow into a wide chimney that opens several inches—occasionally as much as two feet—above the surface at the front entrance. The faster wind over this chimney draws stagnant air out of the burrow, pulling fresh air in through the soil-level entrance. The air is fresh whenever the wind blows, and on the Great Plains that's most of the time. These chimneys of soil draw buffalo as well as air. Bison go to a lot of effort to fill their hair with soil—probably it drives out insects, possibly it keeps them cool. And there's nothing like a good rub on a prairie dog mound.

Then there's that really green grass. Prairie dogs are as much addicted to staying home as bison are to roaming. They both eat grass; but by grazing the same few square yards every day, prairie dogs keep the grass short, and shorter grass is better grass. The closer the blade is to the roots, the higher the percentage of protein and the lower the percentage of cellulose it contains. Closely cropped grass is a necessity for prairie dogs and a treat for bison. So bison spend a lot of time in prairie dog towns, enjoying a snack, rolling on a mound or two, and resting and ruminating.

The relationship appears pretty one-sided so far. Bison mangle the mounds and eat some of the short grass—they are vandals and breadbasket burglars. But bison also bring something to the party—or, more precisely, leave something: grass processed into fertilizer form. It's an excellent source of nitrogen. The longer bison hang around the prairie dog town the more they distribute there, and the more they distribute the greener the grass grows. Of course, buffalo chips don't produce a fertilizing effect as quickly as, say, Miracle-Gro, so the bison are a little like a dinner guest bringing a bottle of wine so new it must be aged a few years to be palatable. Still, prairie dog towns stay put for generation after generation, and buffalo chips are a gift that keeps on giving. Buffalo urine is good for the grass too, and it takes effect right away.

But perhaps the bison's biggest contribution to prairie dog towns is to make them possible. Everywhere but in the western short-grass region, prairie dogs depended on bison to get the grass short enough for them to live there. Prairie dogs won't live in tall grass. Tall grass is less nutritious, and it also hides approaching predators. Where the taller grasses grew, the founding grazers of most prairie dog towns were bison.

A prairie dog town is long-lasting, as are its extended family neighborhoods. Generation after generation of a particular matrilineal group lives in a coterie—two or three closely related adult females and two or three of their yearling offspring. Usually an adult male from another family comes to join them. Every member of each coterie defends the coterie territory against outsiders. And in the spring, when the females begin to give birth to new litters, each mother defends the burrow her litter is in, and several square yards around it, from the other members of the coterie. She feeds on the surface and nurses her pups in her burrow. In a few weeks they are old enough to come to the surface.

The pups that get to the surface are survivors. Some, perhaps many, pups are killed in their nursery, and the most likely suspect is an aunt, an older sister, or their maternal grandmother. John Hoogland has probably spent more time studying prairie dogs than any human alive or dead, and he reckons 20 to 25 percent of all pups are killed by one of these female relatives before they ever see the light of day. The female relatives are fingered as the usual suspects because of a great deal of circumstantial evidence. They have the opportunity—their burrows are closest to the putative victims' and they have the best chance to slip into the burrow while the mother is feeding in the grass. They have the means—the two long, sharp front teeth that are part of the definition of a rodent. And, Hoogland argues, they have a motive—they don't just kill the pups, according to his dark scenario, but they also eat them, thus turning their victims' bodies into milk for their own hungry pups.

Of course it goes against the grain to make the female relatives the prime suspects when a murder may have happened, and there's also the problem of seldom being able to see the corpora delicti. The prosecution has provided a fascinating though rather grisly theory of murder, and the evolutionary logic is impeccable, but so far the evidence is more empty cartridge casings than smoking guns. To be sure, Hoogland caught one female red-handed and -mouthed, bringing another's pup to the surface where she killed and ate it, and with a backhoe he unearthed the partially eaten remains of two other suspected infanticide victims. So it does happen; but the evidence that it happens often is mostly circumstantial. Aunts, sisters, daughters, and mothers did sneak into each other's nursery burrows, spent an average of forty-seven minutes there, and often emerged with an unusual urge to clean something (blood?) off their mouths and forepaws. After such a visit either no litter, or a very small litter, emerged from that burrow. Marauders caught below ground by the mother emerged well roughed up, and those caught trying to enter the burrow were attacked and driven away. The mothers are unlikely to treat their family as dangerous unless they really are.

Analogizing Hoogland's hypothesized infanticide strategy to a crime helps clarify the questions, but we must be clear that the analogy breaks down completely when we come to the issue of guilt. If Hoogland is right, the aunts, sisters and grandmothers are not guilty of anything other than optimizing their evolutionary success. Nature is grandly indifferent to how they go about that—the only thing punished by natural selection is failing to do whatever will give a reproductive advantage.

Given the certainty of some, and the possibility of quite a few, murderous raids on each other's litters, the relationships between the females are remarkably amiable. In fact, the mother and likely murderer of a litter gone missing under suspicious circumstances will be exchanging "kisses" a week later. More surprising still, once they have emerged, pups are likely to enter an aunt's or grandmother's burrow and be allowed to nurse from the owner along with her own pups. By injecting radioactive isotopes into mothers and then checking the scats of pups for particular isotopes, Hoogland showed that at least 68 percent of the pups get at least some milk from a female other than their mother. It makes good evolutionary sense for these mothers to give any milk they can spare to closely related pups—after all, they carry nearly as many of the family genes as their own pups do, so it's almost as good evolutionarily as feeding your own pups. Almost, but not quite. So when you're really short of food it pays to eat your sister's pups, and when you have some milk to spare it pays to feed them. The evolutionary economics are as compelling as the logical conclusion is appalling, and natural selection favors economics over sentiment.

Bison are part blessing, part affliction, but there's no upside to many of the other creatures drawn to prairie dog town. A mountain lion attacking a deer or sheep runs through its repertoire in a few seconds, and its prey's range of defenses is completed as quickly. A prairie dog's encounter with swift aboveground predators such as coyotes and golden eagles is finished in seconds; it either escapes underground or is caught. But rattlesnakes are slow and tunnels are no deterrent. The frantic moments of escaping a coyote or an eagle are replaced by a long, complex dance (it's the right word even though one partner is legless)—sometimes subtle, sometimes brutal, always bespeaking a long, intense relationship that has shaped the behavior of both. Prairie dogs do not go gently into that long stomach. They have a big anti-snake repertoire.

A rattlesnake can't hide on the putting-green surface of a prairie dog town, and it quickly finds itself fang to face with a resident prairie dog. A lone prairie dog may simply retreat, but is more likely to announce the visitor by barking (hence the name "dog") or jump-yipping: flinging itself upright on its hind legs and yipping. Adults bark when the visitor is acutely dangerous—for example, a coyote or golden eagle—and jump-yip when it isn't. They generally jump-yip to announce snakes. A bark usually sends other dogs scurrying for their burrows, but a jump-yip usually draws a crowd. Now it's less clear who the hunter is. The prairie dogs fling soil in the snake's face, turning their backs and kicking with their hind feet. They may dart to its tail, bite deeply into its flesh, then leap away out of range of its answering strike.

But they don't treat all snakes alike. Some are more dangerous than others; prairie dogs' ability to assess just how dangerous has itself been assessed. North America's ground squirrels have attracted a lot of attention from behaviorists. Don Owings and Jim Loughry put a large and a small rattlesnake in a dog town. Adult dogs barked at the large rattler a good deal more than at the small one—a valid assessment, as the small rattler was less dangerous. But dangerousness is relative, not absolute. Rattlesnakes are a deadly threat to a mouse, but not much of a hazard for a moose. Likewise, a given snake isn't equally dangerous to all prairie dogs. The smaller and slower the dog, the more vulnerable, and they behave appropriately: youngsters bark when encountering a snake that draws only jump-yips from adults.

The tip of a harassed snake's tail begins to vibrate, the castanet on it rattling in a buzz that has raised the hair on the back of my neck more than once. The dogs hear it, but unlike me, they probably don't simply get scared—more likely they use it to assess the snake, as their California ground squirrel cousins do. The behaviorist Ron Swaisgood recorded the rattling of a large and a small rattler when each was warm, and again when each was cold. A big snake's rattle is lower pitched, and a warm snake rattles faster. The bigger the snake and the warmer the snake the more dangerous the snake—and the ground squirrels were most intimidated by the playback of the large snake rattling when warm, and least intimidated by the small snake rattling when cold.

After tutoring by prairie dog parents—perhaps even before—prairie dog pups flinch at a rattlesnake's sound. Imagine one setting off to explore the neighborhood, perhaps going down a nearby burrow and hearing from the darkness that rattling warning. Time to retreat.

Yet how odd. Why would a rattlesnake—which takes such risks to get close to young prairie dogs—warn one away? Rattlesnakes bear their young alive and mothers are attentive and fiercely defensive. So it could be a mother defending her vulnerable newborns in a borrowed burrow. But sometimes the rattle is produced by a pseudo-snake, a *very* distant relative that has come to make the same sound: an owl. A burrowing owl, to be exact—a branch of the owl family that lives on the plains as the squirrels do, in burrows. So we call them burrowing owls, but "borrowing owls" would be a better name. They don't dig, they just move into an available burrow and set up housekeeping. Having no sword to rattle, they just rattle. It's as if a mouse being chased by a weasel took on the guise of a hawk.

The first report of burrowing owls making a sound like a rattlesnake rattling was published more than a hundred years ago. But though we knew how it sounded to us—the pulse rate is that of a warm rattlesnake and the pitch that of a big rattlesnake—we didn't know how it sounded to the animals it must be directed at. All ground squirrels will eat eggs or baby birds, so any of them would be part of the target audience. The ecologist Matt Rowe played recordings of an owl's rattle and a rattlesnake's rattle. The squirrels were equally inhibited from entering a sheltering opening.

Despite being closely confined, burrowing owl families are not close-knit. Everybody wanders. Mom and Dad both engage in surreptitious trysts with the neighbors, and females lay eggs in one another's nests. Since nobody can be sure who a particular egg's mother and father are, it may not matter much that the young wander too, leaving home and adding themselves to neighboring broods. At least a few must end up rejoining one or both of their biological parents.

CHAPTER 15

Badgers

I was excited by the anger of the men and excited by seeing the badger, excited by the prospect of killing a varmint. *I* wouldn't kill it, I was too young; but Dad and the neighbor were carrying their irrigation shovels. The neighbor, before he ran out of breath, had been cursing: "Damn badger . . . digging holes all over my pasture."

They weren't running fast. Middle-aged smokers wearing heavy rubber boots and carrying shovels are not fast. But it wasn't far. We'd spotted the badger on the other side of the pasture and started right for it. The badger had seen us too. I'd expected it to run, but instead it dug. A mound of soil grew behind it and it was below ground before we got there.

Boots, shovels, smoker's cough, and all, it couldn't have taken us more than a few minutes to cross the pasture, but the badger was already out of sight. The freshly turned earth was still in motion as more soil was pushed up from below. Too late. But no—we had the shovels. Both men started to dig and in a few minutes had a hole a foot and a half deep. But the badger was still out of sight, and now the tunnel curved round the side of an underground boulder. The shovelers gave up, in frustration and awe. Dad shook his head. "That son of a bitch can dig."

We were in western Montana, mountain country, where a ground squirrel related to the prairie dog lives. Likely the badger was enlarging the ground squirrel's burrow rather than starting a hole from scratch. Still, a badger is several times the size of a squirrel and needs a hole in the ground several times as big. And yes, they can dig. They are, more or less, carnivorous digging machines. Short, powerful limbs. Long, strong claws. A wedge-shaped head. No noticeable neck. Give the above assemblage a motor and an appetite and it will feed itself, digging ground squirrels out of their burrows. Out on the Great Plains that means prairie dogs.

Badgers are members of a family of small carnivores that includes mink, marten, ferrets, weasels, wolverines, otters, and skunks. Every member of the family eats meat and they all have musk glands that emit noxious odors in self-defense. Hence the family name, the Mustelidae. It's a big, diverse, and pretty bloodthirsty family. But though badgers are from a big family, they're not big on family. Behavioral ecology theory predicts most small carnivores will be solitary, and American badgers conform emphatically. (European badgers don't, but that's another genus, another diet—mostly earthworms—and another story.)

The social organization of the American badger fits expectations nicely enough to bring a tear to the eye of even the most jaded behavioral ecology theorist. Behavioral ecology theory strongly parallels economic theory. The things an animal uses—food, shelter, even mates—are resources. Much of an individual animal's behavior and energy is devoted to getting it access to resources. One strategy is to monopolize resources—say, by being territorial. But monopolizing resources doesn't necessarily produce the best ratio of benefits to costs. Monopolizing can be so costly, the theorists point out, that its benefits may not be justified. And so it goes. The animals become a bunch of hairy, feathery, or scaly capitalists investing time, attention, and energy and getting returns in food, safety, and ultimately offspring. These essentially economic models are simple and sensible, and therefore useful, though the practical applications can be challenging. The economist measures both costs and benefits in a common currency—the dollar, euro, or yen. It's reasonably easy to see if you're making a profit or not. Behavioral ecologists can seldom express both costs and benefits in the same coin.

Behavioral ecologists who study nectar-feeding birds are exceptions. They can quantify the benefits of chasing other birds away from a patch of flowers by measuring the extra calories left in those flowers. A few straightforward physiological measurements enable you to say how many calories the bird spends chasing other birds away, and voila! you have costs and benefits in the same coin—calories. You can work out when chasing is profitable and when it is not, then compare what birds actually do with what would be profitable. By and large, nectar-eating birds prove to be good capitalists—they chase when it's profitable and don't when it's not.

However, things are seldom so straightforward. We often find ourselves trying to weigh costs in increased risks of predation against benefits in numbers of females inseminated or eggs incubated. Yet we can still apply the principles in a rough way. Take the economics of a favorite badger food—prairie dogs—as a badger resource. Prairie dogs live in towns. An acre of prairie dog country has either dozens, perhaps hundreds, of prairie dogs living there or no prairie dogs at all. In behavioral eco-economics, the more concentrated the resource, the more "economically defensible" it is: that is, the benefits are likely to exceed the costs and it makes sense to defend it. In badger terms, you may have all you need to eat if you can keep other badgers out cheaply enough. If you don't have to control a long boundary, go far to intercept intruders, or chase them far to get them off your property, then the territory is economically defensible.

In these terms, badger behavior makes sense. The females defend an area large enough to feed themselves, post it with scent marks, and patrol it to keep the food to themselves. For males, it's a little more complicated. Another principle: females go where the food is, males go where the females are. The male is working on two resources: food and females. They need to feed in order to breed, so the boys are after both squirrels and girls. Of course the males are a resource for the females as well, but not one they need to find or compete for. Anyone that happens by will do, more than one usually does happen by, and usually one is enough. The females occupy, relatively exclusively, a bit of land encompassing a bunch of burrows, and the males occupy, less exclusively, an area occupied by a few females.

Territorial songbirds and pronghorn antelope get exclusive occupancy by watching for and flying (or running) at every intruder. Badgers are often underground, and when they are not they are often too low to the ground to see who is there. What works for both sexes is a kind of mutual avoidance society—managing not to be in the same place at the same time. To those with a good nose the landscape is papered with the olfactory equivalent of Post-it notes of the kind that say "From Jack or Zelda" and are date stamped: they show who was where when.

With badgers working at being solitary, prairie dogs with a badger tunneling down the front door don't have to worry about another waiting at the back door. But they might have another worry—a smart, nimble, fast predator that some call God's dog but most call coyote may have teamed up with the digging badger and be waiting to snap up aboveground escapees.

Stories about badgers and coyotes teaming up have long been part of the lore of Native Americans, and for decades some suggestive anecdotes have surfaced in the scientific literature. The ecologist Steve Minta is a field biologist's field biologist. He's able to do field repairs on a motorcycle, overhaul a pickup truck, design a better radio tracking antenna, and spend eighty-plus hours a week watching badgers go about their business. Steve is as flinty-eyed a biologist as you are ever likely to meet. His theme song is "Where's the Data," and all lyrics just repeat the title. So it was a somewhat quirky fate that chose him to document firmly this unlikely partnership and even to quantify its benefits. His badgers lived in western Wyoming, home not to prairie dogs but to their close cousins, Uinta ground squirrels.

Steve implanted a lot of badgers with transmitters so he could find them again, and he did find them often. But sometimes, when he tracked a radio signal to its source, he saw not his badger but a charged-up coyote, ears forward, head cocked, pacing about in an Uinta ground squirrel colony. Coyotes kill badgers, especially young ones, so Steve thought uneasily of the time and trouble that had been involved in trapping a badger without hurting it, and then implanting the transmitter without infecting it. He had to do sterile surgery in field conditions: because badgers have no neck in the usual sense, collars were not an option.

Still, he had come to watch nature take its course, not to mess with it, so he settled down to see what would happen when the badger came to the surface. Several dozen times the animals just went about what proved to be their joint business. The badger would search out a promising burrow and start digging. The coyote would follow and hang around on the surface. The squirrels in a tunnel a badger entered would be eaten if they stayed and eaten if they ran. They seemed more reluctant to run when a coyote was upstairs, so when a coyote waited on top the badger stayed down longer, a pretty sure sign that it was catching more squirrels. The coyote, meanwhile, was catching more than a coyote catches when there is no badger below. Steve could count the victims. So both carnivores benefited. Squirrels caught hell from both directions.

This association could be as impersonal as hawks hovering above a wheat harvester and snatching up the mice exposed in the short stubble the harvester creates when it cuts the wheat. Badgers and coyotes are mortal enemies, after all. But there's more to it. These two have not an impersonal symbiosis but a relationship, and it precedes the partnership.

I would have no idea how to initiate a relationship with the solitary and surly badger, but God's dogs know how, and Steve saw them use their pickup lines several times. The coyote was always the initiator, and often started with the same pitch the family dog uses to get something going—the play bow accompanied by some tail wagging and exaggerated sideways scampering. Pulls you in every time, right? It's not surprising that the often gregarious coyote has the pitch in its repertoire—it has been using it on other coyotes since it was a pup. But it's astonishing that the badger should respond to it. Adult badgers don't invite other badgers to play. Yet the badger responds to the coyote. Both bounce about a bit, then, cautiously, they touch noses. After that, it seems each can safely turn its back to the other and go about its end of their joint business.

While they lasted, these relationships were as transforming as falling in love. Not only did badger and coyote tolerate one another, but when they took a break from hunting they lay down together—sometimes even touching. But these were not long-term commitments. Few such partnerships lasted more than a couple of hours. Afterward, each animal went back to its solo strategy, with the highly specialized badger pursuing its one method—dig and devour.

CHAPTER 16

Coyotes

The versatile coyote, not paired with the badger to hunt squirrels, would probably search for some other kind of food, perhaps with other companions and collaborators. They're able to turn up a meal in more North American settings than any mammal but humans. They'll eat fruit, berries, insects, eggs, and animals as small as mice and as large as (usually young) antelope and sheep, and they'll scavenge winter-killed elk and bison. They even form packs and hunt grown deer in deep snow.

Opportunists that they are, they sometimes include poultry and lambs in their diet and for that we have trapped, shot, and poisoned them relentlessly, but even so we've done them more good than harm. Two hundred years ago, there were no coyotes east of the Mississippi and none in sizable areas west of it. Today they've reached the eastern seaboard and are harvesting housecats in suburban Los Angeles, as well as scavenging winter-killed bison in Yellowstone Park. Their numbers and range are shrinking in Yellowstone, however, and the reason illuminates their recent success elsewhere. There are wolves in Yellowstone again, and wolves kill coyotes. The two species don't just compete for food: wolves search coyotes out, run them down, and rip them apart. It's likely that the single biggest favor we ever did for coyotes was to eradicate wolves.

While it was our actions that created a space for coyotes, it was coyotes' actions that filled it. They are adaptability itself in their social organization as well as their diet. Two local factors seem to determine their size and social life: diet and density. On the desert of the southwestern United States and northern Mexico, coyotes are small and solitary. They wander widely, mostly at night, snapping up mice and moths and daintily removing the fruit from prickly pear cactus in season. They don't need either a large body or help to capture the small things they feed on there, and sharing such small packets of nourishment would leave everybody almost as hungry as before they fed. The smaller their body, the less food they need to find for it. So small and solitary is a good combination. In the snows of Alberta's Rocky Mountains coyotes may form packs, probably of relatives, and kill grown mule deer. The bigger the coyote and the more help it has, the better its chances of killing a deer several times its own size. And if a deer is taken, its flesh will feed a lot of coyotes for a long time. So the coyotes are large there, and move about in packs.

In the mid latitudes their behavior varies—they disperse far from family where food packets are small and coyote density is low, and stay home in a family territory when food packets are small but density is high (too high for them to find an empty space to settle in). Stay-at-homes in such circumstances move about alone because food packets are too small to share. Food packets big enough to share are big enough to be worth stealing, and a group can more effectively defend a parcel than an individual. Family packs at Jackson Hole, Wyoming, get most of their food from really big parcels—elk that die from wounds or old age every winter. The family packs defend these carcasses from other family packs. These packs are organized like wolf packs. One pair breeds and the others, mostly their adult offspring, help raise the pups. It's easy to see why this arrangement works for the breeders—they have both pups and help raising them. It's not so obvious why it works for the grown kids who are giving up reproducing to help raise their brothers and sisters. But brothers and sisters are a lot like your own children—they have many of your genes, so raising them is another, albeit less efficient, way to get your genes into the pool. Besides, it's pretty hard to find a mate and start a family of your own when the food is being guarded by packs.

A few stay in the home pack, waiting to fill an opening at the top. A male may even become his mother's consort. But eventually most grown pups disperse to try their luck elsewhere. Even here the coyote shows how rigorously its behavior has been pruned by natural selection. Males stay on as helpers longer than females, just as evolutionary theory predicts. That prediction comes from a cynical-sounding but fundamental fact encapsulated in a rhyme: "Mama's babies, Daddy's maybes." Mother mammals *know* that the babies they give birth to are relatives. Male mammals never know for sure who fathered their mate's babies, no matter how jealously they have guarded her. Since they're as certain of their relationship to their mother's babies as their sisters are, and less certain of their relationship to their mate's babies, staying and helping has a little bigger payoff for them than it has for their sisters.

The coyote's social flexibility gives them a tremendous potential to reproduce when times are good. Females breed younger and have bigger litters, and before you know it you're up to your hips in coyotes. Terrible as the great slaughter was for the bison in the nineteenth century, it must have been a tremendous boon to coyotes. For a few years there were so many bison carcasses, their tough hides conveniently removed, that there would have been no real competition for food. The coyote population must have exploded.

What was probably the last buffalo hide hunt in Texas, in 1879, killed only twelve buffalo. The hide hunters had put a lot of money up front, and it looked like they would lose their shirts until they noticed the coyotes. The hunters laced some buffalo tallow with strychnine and collected 600 poisoned coyote pelts, from what must have been the densest population of coyotes ever to live in Texas or, for that matter, anywhere on the Great Plains.

Strychnine brings on a terrible death, but those 600 coyotes may have been luckier than most in Texas. The sudden loss of all those buffalo carcasses surely meant slow starvation for many of the coyotes born during the time of plenty. But though this was the last and biggest surge of buffalo carcasses in coyote history, it was far from the first. Severe winters or droughts are a time of great die-offs among bison, and those years supported prolific birthing among scavenging coyotes. And at least some bison died every winter, leaving a frozen meat supply that could last well into spring. Coyotes had many competitors for that meat.

CHAPTER 17

Grizzlies

In parts of the plains coyotes had to compete with one of the most formidable scavengers since *Tyrannosaurus rex*: Griz—the Great Bear. Once known as *Ursus horribilis*, "the horrible bear," it's now known as *Ursus arctos*, "the holarctic bear." It's the brown bear of Europe and Asia, the Kodiak bear of the north Pacific coast, and the grizzly bear of North America's mountains and plains.

Lewis and Clark were the first European Americans to see a coyote, and they were taken by surprise—they hadn't known it occupied the western plains. But they expected to see the great bear, and rather looked forward to it. They knew that the tribesmen along the river—who feared very little—feared the grizzly a lot. Lewis and Clark put this down to the tribesmen's weaker weapons—bows and arrows and lances. Surely a .54-caliber lead ball propelled to full velocity down a 33-inch barrel made to Captain Lewis's order for the expedition would dispatch this bear quite handily. But after one bear swam to an island in the middle of the Missouri River with ten balls from their long rifles in its body, bears began to haunt their days and their dreams as much as they did those of the tribesmen.

Lewis and Clark eventually saw more grizzlies than they wanted to, but griz was not all over the plains. Lewis and Clark were river-bound, searching for a water route to the Pacific. John Fremont took a land route to Oregon in 1842, saddling up in Missouri where the Kansas River joins the Missouri River and heading west. The party did not see a griz until they reached the Wind River Mountains in western Wyoming. Griz was in the mountains and along streams because its food is more concentrated there. The great bear eats mostly vegetation, including a good bit of grass. But its simple stomach can't get much of the good out of grass, though it does better on fruit and roots. Its closest relative is the polar bear, which is almost 100 percent carnivore. Its brown bear brethren along the Pacific coast of Canada and Alaska grow to be much larger because they eat a lot more meat and fish. So from a certain perspective the plains-dwelling grizzly is stunted by its largely vegetarian diet. Little wonder, then, that they so crave meat that they will sometimes spend several ground squirrels' worth of calories digging one out of its burrow. Grizzlies are enthusiastic carnivores, but when it comes to taking on the large hoofed animals on the plains, not very efficient ones.

Stunted they may have been, and rare they certainly were, but grizzlies on the plains loomed large to the people living there, and to everyone who still walks in their range. At least once in a while one will overtake, overcome, and eat a human. Bison have killed a lot of people too, but that's different. They don't eat us, they wouldn't stalk us, and—well—it's just different. Different in a way that makes grizzlies occupy much more space in our imagination than in our mountains or plains. Out on the plains griz was an animal with a polar bear's food preferences and a buffalo's food choices, though it could dig up roots, like prairie turnips, and the occasional ground squirrel. They have been seen killing caribou, elk, musk oxen—even the calves of Alaskan moose. On the plains they could have stumbled over a hiding antelope kit, bowled over a newborn buffalo, surprised a cow helpless in labor. But there's little doubt that they got most of their ungulate meat scavenging in the spring.

Bad winters for buffalo bring good springs for grizzlies. Both go into the winter as fat as frantic fall feeding can make them. The fat will carry both through the winter—though in very different ways. The bear finds a shelter, or makes it—a place in the ground or some protected spot where the snow insulates it against the winter air. Then it turns down its thermostat and goes into a deep sleep, consuming a little fat each day like a slow-burning candle—just enough through the winter to still be alive in the spring. In the spring the thick layer of fat that insulated the bear's core in early winter is nearly gone and the bear wakes up hungry.

Buffalo face winter on their feet: plodding through the deepening snow, sweeping it away from last summer's dry grass with their muzzle, and eating what they uncover to stretch out their dwindling store of fat. In long, cold winters, especially after a dry summer, many buffalo run out of food and fat before spring. And when spring comes the hungry bears feast on buffalo flesh preserved by the cold that killed it.

In the late summer and fall, when the plants are fruiting, grizzlies are almost pure vegetarians. But in the spring, when bears break their winter fast and mothers are rendering their bodies into milk for newborn cubs, a supply of bison as abundant and accessible as a home freezer filled with buffalo steaks must have been a critical element in supporting the plains grizzlies.

Mother bears have a lot to worry about, for though Goldilocks invaded an *Ursus arctos* home, the bear nuclear family is as fictional as Goldilocks herself. Given the choice, a papa bear is much more likely to eat a baby bear than a bowl of porridge, and only the wit, wisdom, and bravery of mama bears keep the cubs alive for the two or three years they need to grow to independence. The great bear's avid appetite for meat may explain part of the male bear's enthusiasm for killing cubs; but females like meat as much, yet leave each other's cubs alone. Indeed, males kill cubs mostly as a reproductive strategy.

This strategy is called *sexually selected infanticide*, and it's perfectly sensible according to the cool—in this case, chilling—arithmetic of natural selection. Each male bear roams over an area where a lot of females live and over which a lot of other male bears also roam. So the chances that any one cub he meets is his are small; it may even have been conceived before he came to the area, so it's much more likely to have been fathered by some other male. Thus he doesn't lose much by killing it, and—here's where the arithmetic gets chilling—he has a good deal to gain. If the female's cubs are killed she will breed again—not one to three years from now, but this season. He has only so many years here or anywhere, and every season counts. So kill he will, unless the mother can prevent him.

Absence is her first line of defense. Female bears usually avoid the habitat where males spend most of their time. But they can't avoid males altogether—when the streams are full of salmon or the river banks are covered with winter-killed bison, they must get a share. The second line of defense is to raise the cost to the male. Mothers have a huge investment in their young and will fight to protect them. Males are bigger and could generally win a full-scale knock-down-and-drag-out fight, but they would pay a price in energy and injury. Often the male will back down and try for the cubs on the cheap, when mother's back is turned.

If he did manage to kill her cubs, the chances are the mother would not carry a grudge, but would soon be receptive to him. Other mothers—lions and langur monkeys—quickly mate with their children's killer, and it makes evolutionary sense. For one thing, the sooner they can get replacement newborns on the ground, the more grandchildren they will have. And, perverse though it may seem, the male would have successfully pulled off an important reproductive ploy that should impress the mother. If sons by him have the same talent it should increase their reproductive success, thus giving her more grandchildren.

CHAPTER 18 Ferrets

A grizzly digging for a prairie dog would be visible to the naked eye a country mile away. A badger doing the same work would be in plain sight for at least 200 yards. But the badger has a slender cousin that slips into a prairie dog tunnel during the night without displacing a spoonful of earth.

No human has ever seen the badger's cousin at work in a burrow that a prairie dog dug. What we can see only in our imagination must also take place in the prairie dog's nightmares. Is the dog awakened in its pitch-black burrow by a presence—sound?—smell?—and so is ready to resist or escape? Or is it taken in its sleep? In either case the unseen presence finds the prairie dog's throat and opens it. The dog's body, denied breath and blood, quickly becomes a meal to be eaten on the spot; or perhaps it is dragged out under the prairie night sky to a nearby burrow, where two or three young nightmares wait to be fed.

You can recognize this nightmare as soon as you see its face. The black-footed ferret's face mask reveals its identify as surely as the Lone Ranger's mask hid his. But rather than being a do-gooder with a gun who drives cutthroats from prairie towns, this masked westerner is itself a cutthroat that invades towns on the prairie and kills and devours the residents—sometimes, in a small town, down to the last dog.

Black-footed ferrets are in the same family as badgers but aren't in the same class as diggers. They can dig a bit when they have to, but they don't have to very often. Long, low, and slender, they find prairie dog tunnels to be just their size. So they simply move into a home a dog has dug, evict or eat any occupants, and snooze away the days when most of the animals that would eat them are active, emerging mostly at night to convert the occupants of the neighboring burrows into entrees. Like other weasels, they eat both summer and winter, so a family of ferrets can make a big dent in the numbers of their immediate neighbors.

When the Grand Duke Alexis safaried on the Great Plains, there were billions of prairie dogs and likely hundreds of thousands of black-footed ferrets. But the species is a victim of its own successes, combined with our excesses. Natural selection molded it, body and behavior, into a prairie dog–killing machine; but in giving the ferret that singular success, natural selection pruned away all its other options. Prairie dog towns are the only communities where it can find work, and we have poisoned and shot so many dogs, and plowed under so many dog towns, that black-tailed prairie dogs will probably soon be listed as an endangered species. The black-footed ferret has become perhaps the rarest and most endangered mammal on earth. And while bison and black-footed ferrets once lived together on most of the vast North American Great Plains, today there isn't even an acre left where they do.

PART SIX

Human and Buffalo

Bison bison *reached its modern form in North America about 5,000 years ago. By then humans had been here at least 5,000 years, during which time they hunted the bison from which today's animal evolved. So small-scale human hunting was one of the forces that made modern bison what they are. But large-scale human hunting was the force that nearly made them extinct. The destruction of the buffalo is a more complex story than the one that we usually hear. That history, together with the bison's hairbreadth escape from extinction, is told in the chapters that follow.*

But we humans sometimes want more from animals than just their hides and meat. We want, in a certain sense, companionship. That desire, and the ways it has connected us to bison, is the subject of the first chapter in this section.

CHAPTER 19

Close Encounters of the Buffalo Kind

Big sniff. Air rushing into vast nostrils an inch from my elbow. Another big sniff, this one above my elbow. The massive head moving up my arm. The dark, quiet eyes and just behind them . . . horns. I'd seen horns like those puncture half an inch of plywood, puncture the rib cage of a bull bison—at a single stroke. Another big sniff at my shoulder, followed immediately by a snort that filled the cab of my pickup with buffalo breath, and the bull turned away.

He left me all atingle, thrilled by the intimacy of the encounter and appalled by my bad judgment in letting it happen. Seemingly peaceful bison have killed many people. I hadn't invited him. It was a hot day, and I had opened my pickup's door to catch a little cool air. I was looking into the distance, binoculars tunneling my vision, and suddenly he was there—close. And I let him get closer.

What is it about wild animals and us? Why does their nearness thrill us? Does the thrill come from conquering our fear of letting them come close, especially if they are dangerous? Some of it, surely. But a part of it is what they do for our egos. I once spent several pleasant weeks near the top of Mount Evans in Colorado chatting with people I saw feeding mountain sheep. Some of these bighorns had become regular roadside beggars, mooching snack foods along the road, often a stone's throw from a sign firmly forbidding visitors to feed wildlife. People gave reasons for tossing or holding out a potato chip to a bighorn. "Better look." "Better picture." But there was more to it than that. When a bighorn plucked a potato chip from between their fingers, they got a better feeling about themselves. They'd had the courage to hold it, and—this is what really matters—the sheep had trusted them enough to take it. Trusted by a wild animal! Nearly everyone told me that animals are better judges of character than people, and they had been judged and found trustworthy by a wild creature. Since animals are more discriminating than people are, their acceptance is more affirming. Either only good people took the winding road up Mount Evans or the trip made good people of them: I never saw a proffered potato chip rejected. Were the sheep recognizing the salt of the earth? More likely they were anticipating the salt of the chips. Still, it was touching to see the two species edge closer to one another—the sheep after the potato chips, the human after affirmation of courage and trustworthiness.

But even after Mount Evans, it's still a thrill for me to be approached by a wild animal, especially a buffalo. And I'm still appalled that I let that bull so close. I went against everything I knew about animals, because I wanted that bull and me to be friends. But not many animals have friends in the human sense, and the humans that think they are friends with such animals are living an illusion—and one that is often dangerous.

I sat with my mother's father after her funeral. He had grown up on a ranch in Oklahoma when it was a territory and became a veterinarian, first working with livestock in South Dakota, then with wild animals for the U.S. Fish and Wildlife Service. He loved animals and he understood them. And he made me understand how my mother came to be killed by one of the family horses. "Joyce never understood horses," he said. "She thought Smoky was her friend—would look out for her." But Smoky had thrown himself backward while she sat in the saddle, and had driven the saddle horn into her heart.

Before you try to be friends with any animal, take a close look at how it treats its other acquaintances, because chances are good that that's how it will treat you. Few animals have much social flexibility. They assign each individual they encounter to one of a small number of categories: members of other species are predators, competitors, or neutral nonentities. Your dog may treat both you and your cat as members of its pack, but treat your neighbor as an intruding member of another pack and your neighbor's cat as prey. The neighbor's cat will likely treat both you and your dog as predators, and your neighbor's dog is likely to treat you as your dog treats its master. Being a neutral nonentity has a lot to recommend it.

If you avoid getting trampled in the rush to be a neutral nonentity, your problems still aren't over. Any change in your behavior, or in the beast's mood, can easily lead to reassignment to a more dangerous status. Stepping toward, reaching out toward, speaking to, can change you into . . . what? A predator to be defended against? A social upstart to be put in place? Hooves and horns are for dealing with both, and when they're wielded effectively they can be lethal. Even if you don't change, the animal's mood may. Cowbirds often attend closely to grazing bison, feeding on the insects that the grazing flushes, and the grazer treats them as neutral nonentities. Yet I've seen an excited bull attack his cowbird contingent, lunging down and slamming a horn into the sod, vainly trying to gore them. The quicker birds easily escaped, preserving both their dignity and their proximity to their insect-flusher.

People aren't that quick. I'm not the only one foolish enough to want to be friends with a buffalo, and many have suffered for it. One proud owner showing off his little herd of pet buffalo was suddenly lifted off his feet when his young bull's horn entered his belly and found solid purchase in his rib cage. The man's luck got better at that point. The upward thrust that gored him tossed him over a fence and at the feet of a visiting veterinarian. The vet kept him alive.

A rancher in Idaho, Dick Clark, raised a bull from a calf, and even when it was full-grown it let him pet it and climb onto its back. It seemed real friendly. Then one day it killed him, mutilated his body, and drove away those who tried to remove the body from the corral. The bull was shot and both bodies removed. The corral that had contained what Clark had believed to be their intimacy now held only their blood. The bull's behavior caused that blood to be spilled, but behind that behavior lay the source of this double tragedy—the man's belief that he and the bull were friends.

A relationship with a buffalo is a dangerous liaison. Bison are immune to our charm, sincerity, personal integrity, and peaceful intentions. We have to learn to value them without believing that they value us. We need to be able to appreciate them without needing to believe that they appreciate us. We need to get it out of our heads that they reciprocate our tender regard or much of anything else. Expecting reciprocity is part of our romantic illusion about other animals. In spite of, or perhaps because of, having learned through experience that members of our own species often don't reciprocate our kindness, we cling to the belief that members of other species will. Surely they're purer, simpler, not devious or ungrateful! Don't bet your life on it.

BUFFALO TAMING

But not everyone who gets close to bison is seeking friendship. Some want to impress their audience, and perhaps themselves, as able to impose their will on a buffalo. That must be at least part of what's going on when people train bison to carry a rider or pull a wagon—seeking respect from the bison and from the beholders of the rider or wagon driver.

Converting a bison to a beast of burden means getting it to walk or run—things it normally does on its own initiative—on command, while wearing a harness or saddle and bridle. The bison resist, and the usual technique for overcoming that resistance is to be forceful; *breaking* them to harness or saddle is the usual term, and it's a good description. Most animals can be socially dominated, at least temporarily and at some stages of their lives. That keeps them from getting maimed or murdered in a battle they can't win. The breaker taps into this adaptation, asserts him- or herself as the dominant in the relationship, and gets submission from the beast. It works. For centuries many of the great systems of agriculture, transportation, military aggression, and defense have depended on this simple technology.

The technology has long been applied mostly to domestics—animals specially selected to respond to humans asserting the dominant role—and has been called *training* (dogs) or *breaking* (horses). Sometimes we do that by breeding for neoteny; that is, selecting for animals that retain some juvenile traits as adults. One such trait is being vulnerable to intimidation. But sometimes humans take on wild animals, often dangerous ones, and the process is called *taming*. That is how we get the biggest, strongest, and smartest of our beasts of burden—working elephants. They've only recently been bred in captivity, so their psychology hasn't been shaped by human goals. They were captured in the wild while still young, then emotionally and physically abused—there's no better way to describe it—until they can be intimidated by a mere human, a creature not much smarter and about 1 percent as strong.

People don't tame bison to get beasts of burden, they tame them to prove either that they are tamable or that somebody has got the stuff to do it—in either case, for an audience. Ron Whitman trains lions and tigers for a living, and when he's working with them the big cats have all his attention: their power and subtlety are his focus, and he appreciates their depth and their many dimensions. But to the audience the cats are mostly background—roaring, stalking, charging danger that Ron challenges and overcomes with a whip and a chair. He is our focus, defeating our fears dressed up as lions and tigers, thrilling us with our own possibilities: he is both surrogate and savior.

The lion tamer's act only works when the audience knows the danger is real. The blood of a tamer spilled anywhere in the world refreshes the power of this ritual—the power of tamers to redeem us, our belief in the tamer, and his belief in himself. Small wonder, then, that when one of two buffalo bulls tamed by A. H. Cole to pull a cart on his Nebraska ranch gored him fatally, Buffalo Jones rushed to his widow with a handsome offer for the killer bull, which he then harnessed to his own cart. Forcing a proven man-killer to pull a cart would impress even Jones's frontier audience. So Jones, who had hunted buffalo for the market and lassoed some of the last wild survivors of the hunt, added a bit to his legend. Riding in that cart had about the same relationship to transportation as riding a unicycle across a high wire without a net has, and had much the same point— to demonstrate that Buffalo Jones had a lot of skill, nerve, and guts and wanted people to know it. Taming means starting with a wild creature and changing its behavior toward at least one human. You tame the aggression in a creature by confronting it—when you overpower it with your own courage and resourcefulness, your audience appreciates what you show them of yourself. You overcome the fear in a creature by your patience and by convincing it of your trustworthiness, and your audience appreciates *those* qualities in you.

The more truly wild the animals, the better all this works for the tamer. But there are practical limits to how much wildness we can deal with. Replace the four plodding oxen drawing a wagon westward in the 1850s with four wild buffalo bulls and it would have been impossible to find the wagons, much less circle them. Teams of the oxen's wild ancestors would have been about as intractable as Buffalo Jones's bulls, but we domesticated them—breeding for generation after generation the more docile, the less aggressive, until we had a creature obedient to commands given in the high, pure voice of a child.

Maleness represents a special challenge to trainers and tamers. Stallions are harder to dominate than mares—the elephant driver perched on a bull may be toppled by the onset of mating urges. But precisely because males represent a special challenge, they also create a special opportunity. Driving a brace of intact bison bulls hitched to a sturdy cart is the stuff of legend. A more cautious approach to managing maleness is to modify it. Consider the bull named Harvey Wallbanger, described in chapter 4. Harvey had been trained to carry a saddle and rider. For years he ran short races against horses and nearly always won, though the horses seem to have been selected to be a bit slow of foot. Harvey was a bull whose maleness had been managed. To be exact, he was a steer. The surgery didn't guarantee his behavior, but it made it more manageable.

Saddling a fast buffalo and racing off into the sunset appeals powerfully to the cowboy in all of us. But the fact that someone has done it, even with a modified male, and lived shouldn't confuse us about bison. They're big and athletic. They have been selected to be submissive when they must, but they have also, especially the bulls, been selected to challenge that status and stage an uprising, violent if need be, against the dominant. After all, subordinate bulls breed little if at all, and selection would soon eliminate a tendency never to try to rise in rank. Artificial selection in domestication, where aggression usually is selected against, has probably reduced it in the animals we normally harness and ride. So domestication and castration have made our large animals more mellow, more subservient, less dangerous. But their wild relatives, however much they look like the domestics, are still wild.

CONFRONTATIONS

Several times I have heard a grunt, the sound of expelled air and rapid hoof beats, and turned to find a mother buffalo charging. Apparently I had just been reclassified from neutral nonentity to predator, or from distant predator to too-close predator. Some authorities recommend "calling the buffalo's bluff" by standing your ground, waving your arms, and shouting. Running away, they say, would encourage further pursuit.

I've always run away. Calling an animal's bluff works only if it is bluffing. If it's not, then you're not even a moving target. I have little confidence in my ability to intimidate an animal that will attack a pack of wolves or a grizzly bear. The cow has always turned away from my rapidly retreating back and returned to her calf. I suspect I owe my unpunctured backside to the wolf's habit of hunting in packs. If the mother chases one wolf very far, she leaves her calf exposed to the others. No profit, then, in running after a diminishing threat. Getting right back to the calf is a good rule of thumb, and mothers seem to follow it.

In fact, in their everyday lives it pays bison to ignore just about anything that stays at a proper distance. The proper (neutral) distance varies from thing to thing and depends on the particular bison's experience with it and momentary mood, so there's no set rule. But more distance is always better than less. A bison that finds itself in more trouble than it can handle with another bison puts some distance between them. What's good enough strategy for bison is good enough strategy for the likes of me.

Americans and this quintessentially American mammal have a future together, both as cohabiting species and as interacting individuals. For both our sakes, we must let the lessons of our pasts together shape the form of our future. Natural selection shaped each bison, blood, bones, and behavior, to inhabit the center of our continent, dealing with the hunger of its predators and the competition of its fellows. As a wild animal it wasn't selected to interact with us and it isn't very flexible. Even the domestication going on in some lines hasn't gone very far. If we are to have a benign relationship with bison, we humans will have to do most of the adjusting. If we can appreciate what they really are, instead of what we want them to be, our future together can be safer and richer for us, and safer and more secure for them.

CHAPTER 20

To Kill a Bison

Be a bison. A bison cow on the run, adrenaline soaring, heart racing, hooves flying. On the run from what? From whatever the bison all around you are running from. Something scared them, and the voice of your species' experience tells you that when others run, safety lies in running with them. Being left behind is dangerous; so you run, simultaneously spurred and reassured by the pounding hooves beside you, before you, behind you. Predators can't come through another bison, so hold your position and run. Just so was safety always secured: safety from lions, from short-faced bears, from wolves. Running with the herd has meant safety for thousands of generations.

But not this time. This time the predator is man, and unlike all the other predators he doesn't want you not to notice him until he is beside you, doesn't want you to lag behind the running herd. So this morning men waved, shouted as they approached the herd and started the cows and young bulls around you running from them; and as the herd ran, more men rose to the left and the right, guiding the running herd down a funnel of strange sounds and movements. This predator has turned your ancestor's hard-earned knowledge of predators to its own advantage. It has turned your best defense into a deadly weapon. Your escaping will be the death of you.

But that's not what your phylogeny whispers, and so you follow the cow in front to . . . Suddenly, where she ran there is only empty ground, and in a moment, not even that. The plain has ended in a stride and you are running in space. Your feet flail at the sky as your body tumbles, then your breath is gone and your ribs and spine are breaking as the cow behind you falls on you, as you fell on the cow in front. Your body becomes part of the maimed mass at the bottom of a cliff.

By now the buffalo chasers—the young men who had stampeded the herd to this spot—stood looking down the precipice at their harvest. The cliff is in what is now southern Alberta, just east of the Canadian Rockies, where the sweep of the northern Great Plains is interrupted by a valley carved by the Oldman River, which still flows there. This drop-off became a buffalo jump more than 5,000 years ago. Buffalo jumps lie like a string of beads just a few miles east of the Rockies, conforming to the land's contours as a necklace conforms to its wearer's bones and flesh. They were made successful not by just the cliffs but also by the plains that drew the bison near them. In many years spring comes early to these plains in the form of a chinook—a hot, dry wind that can clear ten inches of snow overnight. Where such a wind blows, buffalo, hungry and thinned by the draining cold of the months just past, would come to graze.

These warm, late winds are mountain made. The moisture in a mass of moist air flowing up the slope of a mountain precipitates as rain or snow as the air rises higher and gets colder. As it flows down the far side of the mountain, this now dry air compresses and the compression heats it—a lot. It's warmed 5.5 degrees Fahrenheit for each 1,000 feet it descends. After 5,000 feet of this—not an uncommon difference between the top of the Rockies and the plains east of them—the air is 27.5 degrees warmer than it was, and spring can arrive like a fast freight train. A chinook wind in 1900 raised the air temperature in a Montana town 31 degrees in three minutes.

In southern Alberta the chinook starts as a wet wind blowing east from the Pacific Ocean. It loses its moisture as it climbs the Rockies' west slope, then heats as it descends their east slope, becoming a chinook and melting the snow on the plains lying east of them. The grass on the cleared plains drew the bison, and the bison drew the hunters, and the hunters drove the bison over a convenient precipice. The hunters now stand at the edge of the chinook-cleared grassland, looking down at the buffalo they have just driven over the cliff at a place they call "Head Smashed In." Most of the herd will die in a minute or two. Any that survive the fall will be killed by the lances and stone axes of the secret keepers: men waiting below the precipice who believe that any survivors will warn other bison and cause the next hunt to fail. They had a point. Bison that survived such a hunt might be harder to drive next time. And these men were right about the importance of knowledge. Their hunt worked because they had eavesdropped on the whispers from the bison's phylogenic history and turned those whispers to their own advantage. They were astute students of animal behavior.

They used the same knowledge in other buffalo drives at other seasons—into a pit created by dissolving limestone; into a narrow arroyo, a corral the hunters had built of stone and wood; or even, in southern Colorado, into a sand dune. Their understanding showed, too, in the wolfhide hunt. Recipe for a successful hunt: dress and act like your quarry's ancient and mortal enemy, and saunter up very close—being sure to stay in plain sight of your prey. It's not obvious that it will work, but the man wearing the wolf pelt next to his dark skin and approaching the buffalo herd on his hands and knees knows something that isn't obvious. He knows about wolves and buffalo. In some sense of knowing, he can be said to know what buffalo know about wolves. And he knows how to use that buffalo knowledge of the wolf to create a recipe for the buffalo's death.

He knows that only a few interactions between them are a headlong chase—most are a subtle dance. Buffalo and wolves saw a lot of each other and came to know each other well. Bison saw wolves too often to fling themselves into headlong flight every time one appeared. Wolves aren't always hunting—sometimes they're patrolling the boundaries of their territory. Sometimes they're on a social errand. Even if they are hunting, they may not be hunting buffalo, especially if the hunt is undertaken alone. And even when they're hunting buffalo, they may not be dangerous. And even if they are hungry, they don't just fling themselves at the nearest buffalo. An adult cow weighs as much as ten wolves, an adult bull as much as eighteen or even twenty. They are quick, kick like a mule, and can drive a horn through another bull's rib cage with one swing of the head. Wolves, even a pack of them, approaching a herd of bison are not likely to attack a healthy adult. They seek the weak—calves or sick or injured adults. Bulls and cows without calves have little to fear from a single wolf, and not enough reason to spend energy escaping it.

So the healthy adult's best move is to stay put, keep a casual eye on the wolf, and go on grazing, and that's what it usually does—in that sense it "knows" the wolves; and the man on his hands and knees, now very close to the bison, understands the bison's knowledge of wolves. That understanding lets him hide where grass too short to hide a robin stretches for miles. And so hidden, he draws his bow and kills a bison.

The hunter was hidden in plain sight. And though the wolf's skin lay over him, it wasn't really the wolf's skin that hid him, it was the bison's understanding of—really, its adaptation to—the wolf. In this hunt, as in the jump, the hunter not only understands the bison's behavior well enough to predict it, but understands it well enough to turn the prey's strengths to weaknesses, its defenses to vulnerabilities.

The hunter under the robe simulated a wolf passing by in order to be an insignificant part of the bison's world. But sometimes, when the rivers froze, wolves came as a pack and drove the bison onto an expanse of winter ice, where their hard, smooth hooves had little more purchase than a man would find trying to walk on ball bearings. The bison were nearly defenseless, and even the strongest were vulnerable. Sometimes men imitated this wolf tactic—trotting up to the vulnerable beasts and plunging spears into them. The people who practiced this hunt called it a *wolf pound*.

The people's understanding of bison behavior was their most important tool. Their stone weapons and moccasined feet would have been useless without it. Their physical technology didn't allow them to simply overpower bison. But about 300 years ago that suddenly changed. The plains dwellers began to ride horses.

How different the horizon must have seemed to the first mounted Indians. Like many of them, I rode my first horse bareback. I was eight and my gentle gray mare was a year or two older. I fell off a lot at first, but the gray always stopped and waited for my remount. And in a few weeks when I could sit the gray's back the day through, even when she jumped a ditch or galloped, distances suddenly shrank. I visited miles-away neighbors in a western Montana valley called Moiese, where my dad's parents had 120 acres. I looked at the Mission Mountains nearly twenty miles away and knew my pony could get me to them and back in a day. I had seven league boots.

So, too, did the people of the plains who, after walking for some 500 generations, suddenly began to ride . . . starting trips in Montana that took them to Mexico and back. And the mounted man could do another thing—he could outrun a buffalo. Overpowering technology at last.

Native Americans hunted bison in North America on foot for at least ninety-eight centuries. They hunted them on horseback for two centuries. The hunt that the horse made possible differed from the hunts that came before it in a very important way. It overpowered the bison's defenses rather than exploiting them. Now it was the horse that the hunter needed to understand and manipulate, no longer the bison or the wolf.

From above and behind the hunter's horse's head comes a whoop, round heels dig into its ribs, and a quirt lashes its hips. Be that horse for a minute or two, running, muzzle out, ears back as your hooves reach, strike sod, and reach again, one at a time but furiously, as you begin to gallop. Before you other runners, as large as or larger than you, are galloping on cloven hooves. They do not run from the quirt or the heels, they run for their lives and they are fast, but not as fast as you. Your ancestors are the hot-blooded horse breeds—Andalusian, Arabian, barb—spirited horses, always ready to run fast and far. Your species became swift and enduring on these same plains running for their lives, then humans took a hand and selected a line even faster. And so you gain on the running buffalo despite the weight on your back. A rope tied round your lower jaw passes over your neck. Sometimes it pulls to one side or back, but now it lies slack on your neck and from the legs that grip your ribs one or the other knee presses into your left or right side and you are running ten feet or so behind and beside the buffalo.

The rider has worked long and hard to bring you to bearing his weight and accepting the guidance of the rope and his knees. The two of you came together in an intense social pairing of just two individuals, from different and distantly related species. He studied what frightened you, what reassured you, what you liked to eat, and where it felt good to be scratched, but he didn't have to teach you to run: he just released you to do it.

The weight on your back shifts. Your eyes, placed near the sides of your head and giving you a panoramic view, see a narrow, bent piece of wood extend toward the buffalo from above your back. It bends still more, then recovers, as a sharp twang reaches your ears. Now you feel the heels and quirt again and the knees direct you to another buffalo. If you overtake it there could be yet another if there are still buffalo and you can still run, but even a buffalo horse has its limits and there will seldom be more than three buffalo.

The first one you overtook today is staggering now. The arrow was well placed, entering behind the ribs and slicing forward through the diaphragm; that great bellows which contracted and relaxed dozens of times every minute since the bison's birth, filling and emptying its lungs, is leaking. The air into the lungs moves weakly, and the air out is mixed with blood. And suddenly it's as though the bison were running higher and higher, in thinner and thinner air—halfway up the Rocky Mountains, then above them, as if in the Himalayas where no mammal can catch its breath. The brain becomes foggy, the legs unsteady; it staggers, stumbles, falls, and slides to a stop on its side, exhausted—one eye almost touching the prairie that bore its feet and fed its body, the other fixed on a fading sky.

The horseman is beyond the buffalo's fading vision, borne wonderfully fast by his horse's passion for effort and speed. He carries a quirt but seldom uses it. Hot-blooded horses run for running's sake. Necks stretched, ears back, little grunts are seemingly jarred from them by effort and the thud of their hooves on the ground. Could the thrill of that ride ever have paled for the rider? Sitting a horse bareback and feeling the reach and pull of the long muscles his buttocks rest on. Clasping the ribs with his legs. Hearing the little grunts of effort and will. Hearing too the hoofbeats, but they seem to come from outside the two of them, as they have become, for just a bit, one being.

But the business at hand is buffalo killing, and a hunter astride a buffalo horse—a horse that could overtake not just one but two, three, and sometimes four buffalo in a single chase—has to tend to business: get another arrow fitted to the bowstring, select another buffalo to kill, guide the horse to it by the pressure of his knees, shoot for the diaphragm and lungs. All the while he is keeping track of the other hunters riding in a long skirmish line, overtaking and infiltrating the running buffalo. After a few minutes the hunters pull up and turn back to a scattering of dead and dying bison. The hides will be freed from the carcasses, the carcasses dismembered, and everything loaded on packhorses to be taken home.

In time another hunter came who would sustain himself by turning the bison's adaptations into money and nothing but money. He was likely to keep warm in the hair of sheep. His hunting technology also did not exploit the bison's antipredator adaptations. Instead, it exploited a gap in that defense with a new kind of death: death from a distance. It was done with a gun—the Sharps buffalo rifle, usually. It was long and heavy and had to be reloaded after each shot. The cartridge was huge, nearly the length of a man's forefinger. It was .50 caliber, or half an inch in diameter—same as the World War II antiaircraft machine guns. The barrel was long and heavy. Long because the slow-burning black powder needed time to get the bullet up to full speed. Heavy because the weightier the rifle the less it recoiled against the shooter's shoulder, and the more heat it could absorb before becoming too hot to touch.

It would have been nearly useless to the rider of a galloping horse, but it and the stand hunt had evolved together and they suited each other. The stand hunter lay on his belly: it was the buffalo that stood. Buffalo are not the shyest of wild animals, but a standing man would get their attention and probably send them running. The modern species was hunted by humans the whole of its history. So the stand hunter made himself as inconspicuous as possible. Ideally he would have approached the herd with the wind in his face, from the far side of a hill, crawling on his belly to the top. He didn't have to be close. The Sharps shot long and true. If he kept his distance and his luck held, the herd would stand while five, ten, or maybe fifty of its members succumbed to the Sharps. Why did it take so long for them to notice?

J. Wright Mooar was a thinking man, and he had a theory to explain the buffalo's standing rather than running when he began to kill them. He and many other hunters said they shot the leader of the herd first. Since the leader couldn't lead the others away, they simply stood until they fell. Mooar believed he had been blessed with an innate ability to locate the leader. His evidence was that the herd rarely bolted after Mooar killed the bull he had identified as the leader.

Mooar killed thousands of buffalo, and I've killed none, but I still say he had it wrong. His theory of the stand was based on his theory of buffalo sociology, which was wrong—an anthropomorphic projection of his theory of human sociology. It's a bit of a stretch to say a group of bison has a leader, but to the extent that groups have leaders they are cows, not bulls. There was a variant of the stand theory that held the successful strategy was to wound the lead cow—as long as she was on her feet, the herd would hang around. I'm not sold on it, either. Bison don't wait for leadership when they're scared, they just take off. So they must not have been scared—and the question is, why weren't they?

Bison react to things they can see, hear, or smell that seem dangerous. The stand hunter nullified smell by coming from downwind. He gave their eyes little to see by crawling on his belly and staying far away. But then he fired the Sharps and there was a sudden cloud of black smoke on the crest of the hill where he lay, followed in an instant by a boom that would reach any ear within half a mile.

Nearly every wild thing fears the unknown, and flees from it. It's a reasonable rule of thumb. If the unknown thing was not deadly dangerous, you've wasted a bit of time and energy. If it was deadly dangerous, you've saved your life. The possibility of one huge payoff like that justifies a lot of small investments, even though each one is likely to have no payoff at all. Call it the lottery principle of predation prevention. It's sometimes called the life-dinner principle. The predator's dinner is at stake; the prey's life is at stake.

Why then did the herd stand for slaughter? My best guess is that the smoke and the boom were so much like something known that they were taken for it. Nothing in the bison's history had prepared it for the buffalo rifle. Danger always came at them, growing larger in their eye or louder in their ear as it grew closer. The bullet was invisible. But what about the sound? Surely the boom of those black powder rifles filled the ear like a thunderclap. Why didn't the herd flee from the first shot instead of grazing quietly, as they often did, while dozens fell to the rifle one by one?

Perhaps because the rifle filled the ear too much like a thunderclap. Thunderclaps were common where bison evolved, and a grazing herd is as indifferent to them as it is to the rain that usually follows. They simply graze on, their ears filled with thunderclaps and their coats filling with raindrops. Thunderclaps signaled rain, not death, and maybe that's why, while the rifle boomed, they grazed . . . and died.

That gives us a possible explanation for the boom, but leaves us with the smoke—a sudden cloud the size of a garbage can that quickly dissipated. Maybe they appeared to be dust devils. These tiny tornadoes are common on the plains. Heat sets the winds to suddenly spinning, and where they touch the earth they gather dust and dried bits of plants that are airborne for the seconds or minutes the dust devil lives; then they dissipate and drop back to earth. It seems a bit of a stretch, but there must have been something, and it might have been that. Whatever it was, it played a part in the way the southern herd ended: tons of lead fired into thousands of tons of hides, which were then stripped from millions of tons of bloody carcasses—which were left to rot on the ground.

CHAPTER 21

Bison Numbers
Before the Great Slaughter

The difficulty of estimating the bison population of primitive America shrinks in comparison to trying to estimate the numbers in North America just after the Civil War—just before the start of the commercial hide hunt often called "the Great Slaughter." Many forces—horseback hunting, robe trading, habitat degradation—bore down on bison in the eighteenth and nineteenth centuries. Even the weather turned against them. Mary Meagher has pointed out to me that as a short-term climate change known as the little ice age ended in the mid-1800s, rising temperatures and declining precipitation would have lowered the Great Plains' carrying capacity even if hunting pressure hadn't increased (see map 2).

Ernest Seton rose to the challenge of tracking the bison's decline. In a discussion titled "The Dwindle," he estimated that the population had dropped from 60 million to 40 million by 1800, then sank to 20 million by 1850. Apparently giving up at this point, he cites Buffalo Jones's bizarrely arrived-at estimate of 14 million in 1870.

Another great if somewhat eccentric naturalist, William Hornaday, also estimated the 1870 population. He started with Colonel Dodge's 1870 data (see chapter 8). Hornaday asked Dodge to elaborate on the published report of this trip, and Dodge obliged in a letter written the year after his book was published—seven years after the event. The bison were packed in, he said, fifteen or twenty to the acre and, according to what he heard from others, stretched fifty miles long. Hornaday calculated that at fifteen to the acre, a rectangular herd twenty-five miles wide and fifty miles long would contain 12 million bison. Hornaday didn't think the herd was rectangular: "judging from the general principles governing such migrations, it is almost certain that the moving mass advanced in the shape of a wedge, which would make it necessary to deduct about two-thirds from the grand total, which would leave 4,000,000 as our estimate of the actual number of buffaloes in this great herd, which I believe is more likely to be below the truth than above it."

Hornaday apparently assumed that Dodge was traveling through the base of a triangle. According to what I recall of my high school geometry, a triangle with the same base and length as a rectangle contains half the area, not one-third, so Hornaday should have put the herd's number at 6 million rather than 4 million—but it probably doesn't matter much. Waterfowl sometimes migrate in wedges (or at least Vs) for aerodynamic reasons, but that's about it for species that travel in wedge formation, so Hornaday's geometric goof seems biologically irrelevant. Perhaps Hornaday was unconsciously predisposed to this error by his view that the entire North American bison population in 1870, the year before Dodge's ride, was 5.5 million—4 million south of the Platte River and 1.5 million north of it.

Both Hornaday and Seton went along on Dodge's ride, though Hornaday did not attempt to extrapolate to the continental population. By remarkable coincidence, both came up with the same size for Dodge's herd, 4 million individuals, even though they calculated it independently using quite different assumptions about its shape and density.

Why were two of the late nineteenth and early twentieth centuries' greatest naturalists so influenced in their estimate of bison abundance by Dodge's report? Perhaps it was the sense of objectivity that the quantified details lent. The total trip distance was thirty-four miles. It passed through bison for twenty-five of those miles. In his letter to Hornaday, Dodge even estimated the density of bison along his route (fifteen to twenty to the acre). Giving a range of density lent credibility to his estimate, since a single figure obviously would be at least a little wrong. Maybe those numbers made the report so compelling.

Despite their convergence on Dodge's herd, Seton's and Hornaday's estimates for the post–Civil War population of bison diverged widely, and given the dearth of information, there is no basis for choosing between them. Jim Shaw sums it up about as well as anybody can: "One may assume with reasonable certainty that the bison population west of the Mississippi River at the close of the Civil War numbered in the millions, probably in the tens of millions. Any greater accuracy seems impossible."

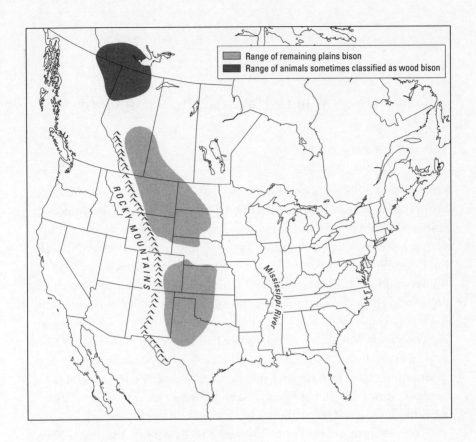

MAP 2. Plains bison distribution, about 1870

CHAPTER 22   Where Have All the Bison Gone?

From tens of millions of buffalo—more than 30 billion pounds of living, breathing bison-mass—to a carpet of whitening bones and a few hundred scattered survivors. From teeming on the Great Plains to nearly extinct within a decade. One of evolution's most abundant massive creatures all but exterminated by a band of malodorous outlaws, supported by a venal military and political establishment whose Indian policy was little short of genocide, who killed buffalo by the millions, took their hides, and left the rest to rot. At least that's the story I grew up on. It's a good western story—guys in black hats, guys in white hats (albeit decorated with a few eagle feathers), and America's natural history despoiled. But, really, the destruction of the buffalo isn't one horrendous story—it's several horrendous stories. The buffalo vanished in different places at different times for different reasons, its slaughter the work of different people.

Many of North America's buffalo were already gone by the time the notorious hide hunt started on the plains. When Europeans came to North America, bison reached the Atlantic coast in the Carolinas and lived in every state east of the Mississippi but Connecticut, Rhode Island, New Hampshire, Vermont, and Maine. The openings in the woods where they grazed owed their existence to active management by Native Americans. These meadows were closely akin to the pastures European settlers would create for *their* cattle, except that the Indians used fires rather than axes. Through fire they managed the habitat to favor the species they preferred to hunt, much as present-day managers might manipulate wetlands to favor ducks, stream temperatures to favor salmon, or rangeland plants to favor deer.

There were still some buffalo left in Kentucky when Daniel Boone explored there in the late 1760s. But their numbers were shrinking in the land behind him. This eastern population withered and vanished under pressures familiar to wildlife biologists today: a combination of mortality—hunters of all races killed them for the table—and shrinking habitat. Europe's diseases spread like a great conflagration among the Native Americans, killing many, shattering communities, disrupting their habitat management. Without regular fire, the woody plants invaded the meadows, displacing the grass. The combination of more firearms and less fire sent the eastern bison population on a long, slow slide to oblivion. By 1833, a little more than 200 years after the Pilgrims landed, there were no bison left east of the Mississippi. But from the Mississippi River to the Rocky Mountains, from the future El Paso to the future Edmonton, a vast sea of grass was still grazed by millions of bison.

Though people hunting the plains on foot had used buffalo meat, hides, bones, and even bladders for millennia, they had probably never diminished bison numbers significantly. Even killing them for the sake of people elsewhere, turning them into items of commerce, was sustainable so long as the trade was small. The Spanish documented trading in buffalo robes and hides on the southern plains in the early 1500s. It had probably started at least 200 years earlier, when bison returned to the southern plains after an 800-year absence. These hides would have been hard come by—probably obtained by hunters on foot driving a few bison into a steep-sided arroyo.

But the horse changed everything. By 1700 mounted Native Americans began to evolve a new culture transported by the horse and supplied by the buffalo—the nomadic hunting bands the world has come to think of as the American Indian lifestyle. Eastern tribes such as the Sioux and the Cheyenne, pushed west by European settlement and pressure from tribes still further east, mounted up and rode onto the plains. Comanches, living a marginalized life at the intersection of the Rockies and the Great Basin, fell upon the horse like the starving upon food and rode onto the southern plains hunting buffalo and growing in numbers and political power. Their nomadic hunting way of life seemed limitless and they rejected the Spanish effort to turn them to settled agriculture. The choice between hunting buffalo and yoking oxen was no choice at all. The bison and the future they would fuel seemed as limitless in time as the view across the southern plains was in space. But the bison were finite; by 1800 the blanket of buffalo that had covered the southern plains was too thin in some years, and the chill of hunger crept through.

The historian Dan Flores has documented and analyzed this decline and its impact on the hunting peoples. Famine was first reported among the Comanches on the southern plains in 1800, less than 100 years after their arrival. They weren't starving in the midst of plenty. Buffalo were sometimes scarce, for on the southern plains they were under a lot of pressure. There were many more people hunting them. They had to compete for grass with feral horses—perhaps 2 million of them. And the plains, and especially the southern plains, had drought years. Drought can reduce the primary productivity of a short-grass prairie by 90 percent, and it takes three to five years for the grass to recover.

And horses brought one more change. They didn't just increase the kill, they made it much more selective. Hunters on foot often had to take potluck so far as age or sex was concerned. But hunters on horseback could choose, and they chose cows. Cows' meat was more edible and their hides—thinner, lighter, more pliable—were more valuable. I know of no accounts of the bull-cow ratios reached in the southern plains, but the historian Francis Parkman hunted along the Arkansas River in southern Colorado in 1846 and saw many more bulls than cows. He attributed the skewed ratio to selective harvest. In the north after forty years of the robe trade General Raynolds (in 1860) and Dr. Hayden (in 1861) independently reported seeing nine or ten bulls for every cow.

It's easy to understand each hunter's choice, but the effect on the bison population was devastating. Selectively hunting cows sent the population in a downward spiral. Bison cows have at most one calf per year, the first born when they are at least three years old. If the calf crop decreases but the wolf population stays the same and continues to kill the same number of calves, then the wolves will take an ever increasing percentage of each year's calf crop. Likewise, the fewer the cows left in a population, the greater the percentage that must be taken to get an equal number of robes each year. Bulls come to predominate in the population, each eating more forage than a cow, while fewer and fewer ever contribute to reproduction.

The pressure on the cows in the north had accelerated in 1820 when competition between two rival Canadian fur-trading companies, Hudson's Bay Company and the North West Company, ended and Hudson's Bay began to worry about competition from the south. John Jacob Astor had established his American Fur Company trading posts in Montana on the upper Missouri River by 1808 and was competing with the Canadian traders for the beaver pelts. To hold their customers the Canadian company began to buy a few buffalo robes and ship them east. Buffalo robes are heavy, and the cost of transportation overland and in canoes down the east-flowing rivers ate up most of the profits, but the trade brought in beaver pelts that might have gone south to Astor. As long as the demand for robes was low and the cost of shipping them high, few were killed for commerce. But Astor's SS (steam ship) *Yellowstone* began to ply the waters of the upper Missouri as far as central Montana in 1831. Suddenly heavy goods could come and go from the heart of buffalo country—buffalo robes could be shipped cheaply. As a large and prosperous middle class rose in North America's eastern cities, demand, prices, and harvest rose too. By 1840 the harvest had risen to 100,000 robes a year.

It was the first linking of steam power and bison, but not the last. The rise of steam and the decline of bison were inextricably linked—in a sense, a large part of the vast herds disappeared into the clouds of industrial steam rising in America and Europe. Fire created bison habitat in the eastern woodlands, and refreshed the prairie grasses further west. But used to turn water to steam, it became the bison's mortal enemy, providing not only cheap robe transportation but also both means and motive for the industrial hide hunt of the 1870s.

In Canada the trade robes were supplied by the people of the prairie. Southern Alberta, Saskatchewan, and Manitoba contained some 200,000 square miles of buffalo country, and hundreds of thousands, perhaps millions, of bison once ranged there. In southern Alberta the Blackfoot tribe had hunted bison for thousands of years. Now hunting horseback, they killed more and traded the extra robes. But most of the robes were supplied by the Métis.

Several generations before, European men hired to move merchandise and furs for Hudson's Bay Company had left its service and, with their Native American wives (mostly Crees), had established a community called *les gens libres*—"the free men." In time their descendants would be recognized by the Canadian government as a Native American tribe called "Métis." The Métis lived as other tribes did, by hunting, trapping, and trading, but with their broader background they readily took advantage of entrepreneurial opportunities. When Hudson's Bay Company began to buy buffalo robes near present-day Edmonton in 1820, the Métis were ready to supply them.

They lived in a wide band across Canada from Winnipeg to the Canadian Rockies. In Alberta they interacted with the Blackfoot Confederacy on a line roughly 100 miles north of the Montana border, but they roamed south to and sometimes below the North Dakota border. They had conducted a small commercial buffalo hunt in summer to provide meat and pemmican to Hudson's Bay Company's workers for years, with a winter hunt for robes for their own use. Now they added a winter hunt for robes to trade. The trade robe hunt was small for some two decades, but by 1840 demand and prices had risen.

The Métis hunts dwarfed any others ever seen on the grasslands. Their southernmost hunting ground was along the Red River where it forms the boundary between North Dakota and Minnesota. Alexander Ross, a contemporary Red River historian, witnessed the Red River hunt of 1840. The hunters were a combined force of Métis and Scottish settlers from the site of present-day Winnipeg. Neither the Big Fifty Sharps rifle nor the stand hunt—the hunter lying on the ground and shooting members of a herd hundreds of yards away as long as they would stand—had yet been invented. This was a horseback hunt, and the Scots had originally been hopeless at it. Their horses had been too slow to overtake a running buffalo, and the riders were not skillful enough to kill from a running horse. Now, better mounted and more skilled, they participated fully.

The hunting party moved like an army. Twelve hundred and ten high-wheeled Red River carts transported 1,630 people accompanied by 542 dogs. (How did Ross get such an exact count of the dogs?) The hunting camp was a small, self-governing city. The first business of the hunt was to elect the chief captain, who would hold office until the next hunt. Ten captains subordinate to him were elected next and each of them appointed ten soldiers. All were empowered to enforce the rules the chief captain issued. These were rules of, by, and for buffalo hunters, and most were to discipline the hunt—to keep the hunters together and synchronized. Repeated violations would be punished by flogging.

Ross described a day's hunt. The scouts located a large herd in the morning. The four hundred hunters mounted together, trotted their horses close to the buffalo together, and galloped for the kill together. When the dust raised by the galloping beasts settled, he could see that the tactics had produced 1,375 tongues.

Robes flowing out, trade goods flowing in—and no longer just the old reliable firearms and firewater, but goods once considered luxuries. In the early 1870s the Blackfeet in the Bow River Valley in southern Alberta were trading robes for flour and molasses brought north from Fort Benton, Montana, on the Missouri River.

In 1875 the robe price began to sag under the weight of more and more robes, but the luxuries had become necessities and the Métis intensified the harvest to recapture their lost income. Soon the bison population was plunging—they were commercially extinct in the northern and eastern portions of their Canadian range by 1876. In 1879 the last significant population was in southern Alberta, in the Bow River Valley and the Cypress Hills. Before the year was out, only a few stragglers remained in Canada and many of the Canadian hunters had moved south into Montana. There they found themselves competing with buffalo hunters who had just migrated to Montana from the plains further south. The story of those hunters, and the buffalo they hunted, is the story of the industrial hide hunt—often called "the Great Slaughter."

In the 1860s steam-powered locomotives began to penetrate the midsection of the United States. Swiftly traveling rivers of steel, they transformed transportation, life, and economics on the Great Plains. In 1869 the Union Pacific completed its track. People and goods could travel from San Francisco to New York in eight days, though at times they had to creep through buffalo herds miles wide. Usually some of the passengers banged away at the buffalo with weapons ranging from heavy rifles to pocket pistols. Ever entrepreneurial, the Union Pacific scheduled special excursion trains into buffalo country whose passengers' only luggage was arms and ammunition and whose highest hope was for a trackside encounter with a bison herd that could be weighed down with some lead bullets.

Pointless as these impromptu or scheduled steam-powered shooting sprees were, they probably had little or no effect on the bison population, even locally. Through the 1860s the only market for the southern bison was for a little meat. Their coats, even in winter, lacked the dense hair that makes a premium robe. But a new link between bison and steam emerged. Boxcars could haul buffalo hides from the grasslands to the East Coast in a fraction of the time of wagon train transportation, at a fraction of the cost. Initially the innovation had no effect, because buffalo hides processed like cattle hides made poor leather. But about 1870 things changed for the bison and the Comanches and Cheyenne; first in Germany and then in the United States, tanners learned to make useful leather from buffalo hides. When the bison's hides could be tanned, their goose was cooked.

Leather was in great demand just then. The industrial revolution was firing up its motive force—steam engines—in eastern North America and Europe. These industrial engines used steam to spin a wide, flat steel wheel. A long belt fitted over both the steam engine's flat wheel and another flat wheel on the industrial machine to be powered, much like the belt with which a modern automobile engine drives its fan. Thus industrial belting was a critical link, and most industrial belts of the time were made of leather. Buffalo hides made good industrial belting. The steam engine powered not only industrial machines but an industrial hunt that devastated the plains bison populations that still survived south of the Arkansas River.

The rise of steam and the decline of bison were also linked by the Union Pacific's locomotives. The buffalo south of the Arkansas River fell to a kind of military-industrial complex. The military part was the intractable war with the plains Indians. Japanese tradition says the world rests on the back of a turtle; the plains Indians' world rested on the back of the buffalo. Many soldiers would have been glad to see the plains rid of bison, but the military didn't slaughter the southern herd—an industry did.

In 1872 a teenage meat hunter named J. Wright Mooar sent fifty-seven hides from Kansas City to his brother John in New York. John immediately sold the hides for $3.50 apiece. When the tanners who had bought them urgently requested more hides, John quit his jewelry business and went west to hunt with Wright. They were a start-up company in an exploding industry. It was a hide rush, similar in many ways to the gold rush to California two decades earlier. Most of the rushers were the currently unemployed. Some were restless young men seeking adventure and wealth. Others were Civil War veterans from both sides seeking wealth, or at least a living. A few were older men drifting west, away from abandoned responsibilities, too much whiskey drunk, too many crimes committed—just doing what came to hand, and hoping to get by for a while.

After a season of eating buffalo three times a day, sleeping on the prairie soil, and wrestling with sometimes decaying hides, these hunters took on the complexion of the soil and the odor of the hides. And while these acquired attributes made the men unacceptable company even by frontier standards, they could not be depended on to deter angry Indians. At first the Comanches and Cheyenne had participated in the hide trade. But when they found themselves competing for buffalo on hunting grounds they believed were theirs alone and foresaw the consequences of the industrial hunt, they began to hunt the hunters, collecting scalps instead of hides.

Most of the aspiring industrialists went on to something else—anything else—after a few weeks or months. But, like the gold rush, a few made some money. Levi Strauss got rich in San Francisco making copper-riveted pants for the Forty-Niners. Christian Sharp manufactured the Levis of the southern plains hunt—a .50-caliber rifle designed to kill buffalo from afar. The hunters dubbed it the Big Fifty and it was their favorite weapon by a long shot, even though it cost $125 (equivalent to about $2,000 today). Then there were the middlemen, the retailers who put these goods into the hands of the rushers, and those who handled gold dust or hides—for a fairly sure profit.

Finally there were those few that made a success of the hunt itself. They were a minority, but they took a lot of hides. By 1875 John and Wright Mooar were hunting the plains like whalers, with a fleet of sixteen prairie schooners to move supplies out and hides back. They hauled hundreds of pounds of lead to make bullets and barrels of powder to propel them—ammunition for a small armory of Big Fifty Sharps rifles. The Mooars did nearly all the shooting. The skinners were mostly outlaws that Wright hired in Fort Griffin and in the backcountry. Wright preferred outlaws: they seldom complained about working conditions and didn't hunger to get to town. The Mooars took 4,500 hides that season.

There were critics of the slaughter. There were objections to the waste, the destruction of the Indians' livelihood, the farce of pretending that killing buffalo was sport hunting and therefore entitled to the recognition that was sportsmanship's due. A lieutenant colonel wrote from the frontier to condemn the lack of sportsmanship in the killing of the animals and the deprivation inflicted on the Indians. Citizens protested and organized, editors wrote condemnations, legislation prohibiting the slaughter was proposed and sometimes even passed. Congress passed legislation to limit the hunt, but President Grant pocket-vetoed it. Editorials and letters and laws are printed on paper, and paper can't bulletproof a buffalo.

The military supported the industry passively. Though it gave the hunters powder and lead, its role has often been inflated. A frequently told story goes as follows: In 1875, when some in the Texas legislature tried to limit the destruction, General Philip Sheridan intervened on behalf of the hunters, urging the legislature to "let them kill, skin, and sell until the buffaloes are exterminated." According to this account, Sheridan justified the extermination of the bison as a means to eliminate an enemy army's base of supplies.

It's a compelling tale but historians can find no evidence that the Texas legislature ever considered a ban on hunting bison or that Sheridan ever addressed it on any subject. The only source is an ex–hide hunter named Cook writing thirty years later, defending himself against criticism of his role by pointing out it seemed right at the time and adding as justification, "It was even said that General Sheridan . . . "

Cook's eagerness to spread the blame is understandable. The hide hunters fairly quickly became objects of opprobrium for what was coming to be seen as a national tragedy—the squandering of a natural resource that would never be seen again. It's by no means certain that many bison would have survived for long even if the hide hunt hadn't taken place, but the hunt was clearly the coup de grâce that closed the bison chapter in a very few years.

In 1877 Joe McComb's wagon train hauled a thousand pounds of lead and five kegs of powder out and 9,700 hides back. But oversupply depressed prices and he got only a dollar a hide. By 1879 the southern herd was gone. The Mooars had already quit the business and invested their profits in ranching. Jack Harris, a die-hard hunter and scout, formed a buffalo hunting party in November 1879. After a long search of the Texas plains they found twenty-two bison, killed twelve, laced the carcasses with strychnine, and eventually collected the hides of more than 600 poisoned coyotes. Each buffalo hide brought five dollars; each coyote hide fifty cents. The hunt was over: the southern herd just a memory—memorialized by bleaching bones. J. Wright Mooar's first hunt was only eight years before Jack Harris's last.

Suddenly the southern plains had a surplus of Big Fifty shooters and buffalo skinners. Some stayed in business by migrating north. The last wild bison in North America ranged in the eastern half of Montana, but in 1879 the hide hunters from the south and the Métis from the north began to chew up these survivors of the robe trade, hunting them for hides in the summer as well as robes in the winter.

Although a few of the actors were the same, commercial hunting in Montana was a very different scene. In the south the Comanches and Cheyenne had tried to stop the commercial hunt; in Montana the Blackfoot Confederacy participated in it—and had done so for decades. The northern hunt was practiced and efficient: very few hides or robes went to waste, although more than 99 percent of the meat did. The great wagon trains that supported hide hunting in the southern United States and the Métis hunt in Canada were missing in Montana. Parties were small— sometimes just one man hunting for a few robes in the winter to supplement wages from his day job. L. A. Huffman was one of these, but his day job was photographer and he took his camera into the field, producing most of the photographic record we have not only of a bison hunt but of wild bison living anywhere on the Great Plains.

The hunt accelerated from 1879, with the peak kill in the winter of 1881–82, when deep snow made the buffalo especially vulnerable. The next winter, 1882–83, saw the last significant harvest. The American bison was commercially, and almost biologically, extinct.

CHAPTER 23

Attitudes

Today the hide hunt and the hide hunters seem utterly foreign to most of us. We wonder what they could have been thinking—no, feeling—that allowed them to kill and waste on that scale. We wonder what kind of people they were, what kind of attitudes they could have had to be willing to do what they did. Yet while the hide hunt was under way, the hunters got some good press. In December 1874 *Harper's Weekly* published an engraving by Paul Frenzeny and Jules Tavernier, who had seen the hunt. It shows a handsome, clean-cut young man standing erect. In one hand he holds aloft one end of a bloodless hide he is stripping from a peaceful-looking bison corpse. In the other hand he holds a bloodless knife. The text reads, "The hides are taken off with great skill and wonderful quickness, loaded on a wagon . . . and carried to the hunter's camp." But before long, many Americans began to mourn the bison's passing, to see the hunt as a sort of murder, and to see the hide hunters as quasi-criminals. And that's about how we see them today—greasy, dirty, conscienceless killers with the blood of millions of buffalo on their hands.

The hide hunters have achieved the anomalous status of despised frontiersmen. Compare their image, as the historian Elliot West has done, with that of the mountain men who went west a couple of generations earlier to trap beaver. They too drove an abundant and ecologically important native mammal nearly to extinction in order to turn its hide into an item of international commerce. If any ever repented of his deeds, we haven't heard of it. Yet our picture of him is not of a gray and greasy scoundrel bent to his butchery over a half-stripped carcass. The mountain man in our mind is Jeremiah Johnson as portrayed by Robert Redford. With classic cheekbones, brilliant highlights in his freshly shampooed hair, and thirty-two perfect teeth, he leans lightly on his Hawken rifle—which somehow seems more accessory than weapon.

The differences were not in their deeds but in the context in which they did them. As the historian Andrew Isenberg points out, when the hide hunt began, the times were changing. The American Society for the Prevention of Cruelty to Animals had been established in the urban East in 1866; and as the hide hunt horror stories came back from the West, the ASPCA for the first time extended its concern beyond pets and domestic animals to wild animals—it declared the hide hunt "cruel." Moreover, the nation had just fought a terrible civil war, largely to end slavery. Humane concern was high and the plight of the Indians starving in the hide hunters' wake troubled many. At bottom the difference between the mountain men's and the hide hunters' images reflected a change in public attitudes.

Public policy toward wilderness and wild things implements attitudes—our private values. The value we put on animals feels so natural and right and inevitable that it's a shock when we first learn that others feel differently about an animal's death. Decades ago I explained to a friend that the state of California was shooting burros in Death Valley because they were driving the desert bighorn sheep to extinction. The bighorn were here before we Europeans came and brought burros, I said, so saving them was obviously the right thing to do. He bridled. He pointed out that there were more burros than bighorns. It would cost more lives to save the bighorns than to let them go extinct. I bridled. He stared. I stared. Not a communication problem—we understood each other perfectly. And we were appalled.

There is no more dramatic illustration of such differences and their consequences than the public policy debate in the 1870s about fate of the bison. At one level it was about consequences: impact on Native Americans, impact on the bison as a natural resource, proposed legislation. But the debate drew on, and illustrated, basic attitudes toward wildlife in general and bison in particular. Then as now, attitudes were mixed. Listen to the historian Francis Parkman, who traveled on the Great Plains in the 1840s:

Except an elephant, I have seen no animal that can surpass a buffalo bull in size and strength, and the world may be searched in vain to find anything of a more ugly and ferocious aspect. At the first sight of him every feeling of sympathy vanishes; no man who had not experienced it, can understand with what keen relish one inflicts his death wound, with what profound contentment he beholds his fall.

It seems likely that shooting an elephant would have made Parkman even more content. Yet shooting an elephant brought George Orwell profound discontent. As a young policeman in Burma, he was caught in a web of circumstance that forced him to shoot a working bull:

> I did not want to shoot the elephant. I watched him beating his bunch of grass against his knees with that preoccupied grandmotherly air that elephants have. It seemed to me it would be murder to shoot him. At that age I was not squeamish about killing animals, but I had never shot an elephant and never wanted to. (Somehow it always seems worse to kill a large animal.)

Neither Parkman nor Orwell acknowledge that anyone could feel differently than they do. What a debate these two articulate men might have conducted, pressing each other with their firmly held but opposite views. But it's very unlikely that either would have changed his mind.

Animals' lives and deaths mean different things to different people. And not just others in a faraway place or long ago time, but next door— today. Parkman's and Orwell's attitudes, and my friend's and mine, are alive and thriving now, as they have been for centuries. When it comes to the meaning of animals to people, contemporary Americans give new meaning to the word *diversity*. This diversity of attitudes is rooted in our history as a people, and it underlay the debate over the fate of the bison.

Parkman exulted in killing a bull buffalo because of its strength. It was a mighty enemy slain, a Goliath brought down, a threat ended. Exterminating buffalo would be like developing a vaccine against polio or eradicating smallpox. Some saw, and still see, bison not as dangerous but merely obnoxious. Last year a stranger remarked to me, "They're stupid and they stink." *Be still my hackles!* He didn't mind at all that our forebears left us so few buffalo—such benighted beasts had no claim on continued existence.

Most hide hunters had a utilitarian perspective: a buffalo has a hide, I can sell the hide. To get the hide I shoot the buffalo. Frank Mayer knew there were millions of buffalo on the plains and he needed money. Many rural Americans still feel that way about animals. When I was a boy my father paid me a two-cent bounty for each dead ground squirrel, and the county paid me a two-dollar bounty for each coyote. I was a bit squeamish about the killing—perhaps the hide hunters were too; but the money was good, the varmints were bad, maneuvering for the kill with gun or trap was exciting, and succeeding was an ego booster—an accomplishment.

Shooters who rode cheap trains or mounted expensive safaris to kill buffalo were looking for a chance to show off a winner's qualities—the courage and skill to kill. But not everyone was convinced the sport hunter's pride in killing buffalo was justified. Basic marksmanship was the only skill required; and as for courage, the Texas longhorn cattle of the day were probably more dangerous. And others objected on other grounds. Editorials in the 1870s condemned the "wanton cruelty" of the hunt on humane grounds. The bison were defenseless. That rings a bell for us to the degree to which we identify emotionally with the animal—thus the degree to which the good or bad things that happen to it happen to us. It's the feeling beneath the angry question/accusation "How'd *you* like it if somebody did that to you?" That part of us feels their pain.

A different attitude was revealed in the words and work of founders of the American Bison Society: an affection and respect for animals as natural phenomena. To them bison were interesting, inspiring, valued as a part and product of nature and thus for its own sake and the sake of the natural world.

Most of us live by some mix of attitudes toward animals, with the one that dominates at any moment strongly influenced by the species at issue. I could pull the legs off a spider for a good reason, but I couldn't pull the legs off a spider monkey for any reason. A good number of Americans despise wolves but love dogs, even though dogs are simply domesticated wolves and look and behave much like them. We all know, and many of us are, people who would abhor eating Bambi but enjoy veal.

Wildlife management, whether through neglect or protection, always has been and always will be the public implementation of private attitudes, with politics deciding whose values get implemented in each particular case. Buffalo were carried to near extinction by fear and utility. Then other attitudes saved them from total extinction and gave the species its present hoofhold on the future. But its hoofhold is just that, and its future will be decided by the private attitudes of the future and by the politics that determine which attitudes are implemented. In America today, all wild things and wilderness always hang by the slender thread of individuals giving a damn.

CHAPTER 24 Conservation

Then and Now

On a quiet Sunday afternoon you could hear Andy Hodge's guns for miles around the National Bison Range headquarters house. He'd practice with the .38 pistols for a while, then the 30.06 rifle, then the pistols again. Andy was good with those guns, and seemed ready to use them. People said he'd killed a man in Missouri before he came west to Montana. He wore an army overcoat summer and winter, and carried his Colt .38-caliber automatics in the large patch pockets. When he drove his Model T he put one of the pistols on the seat beside him. My father, who as a young man had worked for Andy, used to ride with him sometimes and always saw a pistol there. Dad loved to recall the night Andy ran out of gas. Several men were sleeping in a nearby bunkhouse and Andy knew they had gas, but they didn't answer his shouts. So he fired several bullets through the bunkhouse wall just below the roof. "Them guys tumbled right out," Dad would chortle, "and old Andy got his gas in a hurry!"

Andy was the first superintendent of the National Bison Range, the first refuge created specifically to prevent wild bison from becoming extinct. He had been the toughest man on the crew that built the fence around the range in 1909, and so had become the foreman. Conservation then was mostly a matter of making sure the few animals that remained were not lost to poachers. Protecting bison was a dangerous job, best done by a dangerous man. Andy was just about perfect. I don't know how much of Andy's fierce protection of the bison was motivated by defending his own image as a tough guy. I do know the bison weren't poached.

The bison Andy protected were descended from a few calves—survivors of a Blackfoot Indian hunt sometime during the winter of 1872–73 on the plains half of Montana, just east of the Rocky Mountains. The hide hunt was then just building a head of steam in Kansas. Samuel Walking

Coyote was a Pend d'Oreille from west of the Rockies, living and hunting with the Blackfeet. The hunt orphaned some calves. Motherless bison calves often attached themselves to anything big and moving; when eight calves wandered into Walking Coyote's camp, he saw in them the solution to a marital problem.

While living in the Flathead valley, Walking Coyote had married a Flathead woman who had converted to Catholicism. The church had sanctioned the marriage. He had left his Flathead wife in the valley, and while living with the Blackfeet he had married a Blackfoot woman. He wasn't looking forward to returning to the Flathead valley and the church's condemnation, but he believed that a handsome gift to the Jesuit fathers would smooth things over. In the spring of 1873 he returned to the Flathead.

The calves followed his horse through the mountains to the valley. Six of them survived the trip, but didn't propitiate the Jesuits. The fathers, either ignorant or intolerant of native customs, rejected the calves and had the Indian police beat Walking Coyote. He kept the calves, they began to have calves, and by 1884 he had thirteen bison. Two local ranchers, Charles Allard and Michael Pablo, bought the herd for $2,000 in gold— probably the first time in Montana history that a bison was worth more alive than dead. This was the year the Lacey Yellowstone Protection bill became law, finally offering some real protection to the last twenty-five wild bison in the United States.

Samuel Walking Coyote wasn't the first man to keep calves and sell bison. As early as 1870 a few men had captured a few calves and let them breed. Small private herds were developing in Texas, Kansas, Nebraska, and South Dakota. They were financed by their owners' interest and by occasional sales to zoos and other fanciers. The owners were businessmen—entrepreneurs with the vision to see that living bison might be even more valuable than dead ones. But while privately owned bison increased in number, publicly owned bison dwindled.

In those days, private ownership was the bison's best chance for survival. Despite legislation proposed—and sometimes passed—to protect them, publicly owned bison had almost no real protection. Even after Yellowstone Park, the wild bison's last refuge, was established in 1872, poaching continued; the 200 head there when the park opened were steadily depleted despite the efforts of the U.S. Cavalry. Since the army could not legally arrest poachers, the soldiers could only make them uncomfortable. When they caught a poacher they confiscated his weapons and horse and escorted him out of the park on foot—always to the most distant boundary. After walking as much as fifty miles the poacher would be interviewed by a tough sergeant, who put on leather gloves beforehand to protect his knuckles.

With the Lacey Yellowstone Protection bill the federal government had at last put itself in a position to protect its bison, and not a day too soon. It owned some twenty-five in Yellowstone and a handful in the National Zoological Park in Washington, D.C.—the public had a very fragile hold on the remnants of this once teeming species. Things were perilous and remained so for a decade until the American Bison Society was founded in 1905. The founders were shakers and movers, in touch with money and power. The politically powerful and savvy chief taxidermist from the U.S. National Museum in Washington, William T. Hornaday, was its president. Teddy Roosevelt, then president of the United States, accepted appointment as honorary president of the American Bison Society.

Meanwhile, the herd that Samuel Walking Coyote's calves had founded grew, and the new owners bought twenty-six more. By 1895 the Pablo-Allard herd was 300 strong and growing. Allard died that year and his heirs sold their half of the herd to several private parties. Michael Pablo continued to graze his half in the valley on Flathead reservation land, and by 1906 he had some 600 head. Anticipating the end of the open range in the valley he offered the bison to the U.S. government, but Congress dithered. The Canadian government didn't dither, and Pablo sold all his animals to Canada for $200 per head. In the six years it took to round them up and ship them to Canada they continued to breed and Pablo eventually sent 700 head. The Flathead Valley, which in 1906 held most of the bison in the United States, now held only a few dozen—derived from animals that Allard's heirs had sold to local ranchers.

The American Bison Society concentrated on establishing and protecting federal herds of bison, and it moved its agenda quickly. While Pablo was shipping his herd to Canada, the society had proposed establishment of a National Bison Range in Montana to increase the quantity and quality of public herds. It sponsored a survey of Montana for a site specifically for a bison refuge. Professor Morton J. Elrod of what is now the University of Montana at Missoula recommended twenty-eight square miles at the southern end of the Flathead Valley where Samuel Walking Coyote's eight calves had founded what became the Pablo-Allard herd. The site was part of the Flathead Indian Reservation. (The people are actually Salish.) The Dawes Act of 1887 had reduced the tribe's claim to land on the reservation to little more than first choice of sites for 160-acre homesteads. Land remaining after the tribe's members had selected their homesteads was "surplus" and the federal government could buy and resell it, or open it to homesteading by non-Indians. The site Professor Elrod selected was poor farmland and so largely surplus.

It wasn't a Great Plains grassland. It was Palouse prairie—like the grasslands of southeastern Washington and southwestern Idaho. And buffalo hadn't lived there, at least in anyone's memory. But it was available, the Pablo-Allard herd had shown that buffalo could thrive there, and Professor Elrod, who lived less than forty miles away, had known it and recommended it.

Congress was embarrassed by Canada's coup in buying the Pablo herd. In 1908 it passed, and President Theodore Roosevelt signed, a bill authorizing purchase of this surplus land. The range was fenced in 1909 and thirty-seven bison arrived, all but one purchased from a fragment of Allard's herd sold locally by his heirs. But those bison were not provided by Congress, whose embarrassment extended only to putting up the money for the land, not to paying for acquiring animals. It suggested that the American Bison Society provide the bison, and the society solicited money from the public. When I was growing up there was a story in circulation that schoolchildren all over the country had raised the money by contributing pennies. I loved the story—still love it—but a lot of the money came from a few deep pockets and little if any came from children. New York led the pack. Charles Senff contributed $1,000—10 percent of the total. Add William Clyd's $500, and $250 each from William Sloane and Andrew Carnegie, and four big donors are responsible for 20 percent of the total. Eighty-three other New Yorkers contributed $2,725. Four thousand seven hundred and twenty-five dollars—nearly half the $10,560.50

total—thus came from one state on the eastern seaboard. Money came in from every state except four: Kansas, North and South Dakota, and Texas, the heartland of the original buffalo range.

While the National Bison Range's herd was the first established with the conscious goal of establishing separate populations instead of just one to guard against extinction, small sets of animals had been gathered elsewhere in the United States. The American Bison Society was instrumental in persuading Congress to fence 8,000 acres of the National Wichita Forest Reserve in Oklahoma in 1907. New York City zoos had two herds, one prospering so vigorously that fifteen animals were sent by Hornaday to the Wichita Mountains Refuge to found a herd there. That herd, with the addition of a few animals from rancher Charles Goodnight's Texas herd, eventually prospered. The American Bison Society also furnished six bison to start a herd at Fort Niobrara, Nebraska, in 1913 and fourteen the same year to start a herd in what is now Wind Cave National Park in South Dakota. All these animals that served to found public herds came from private herds.

At the National Bison Range, success in the first stage of conserving bison eventually brought the welcome problem of too many bison. Too many, that is, for the 18,000 acres of grassland that had been set aside for them. The range started to die. Biologists in the U.S. Fish and Wildlife Service decreed that the herd must be cropped. Andy seemed to see no important distinction between the poachers and these distant bureaucratic biologists. Nobody would kill any bison on his watch. The Fish and Wildlife Service insisted, and Andy left.

Andy's successors were a different breed. Andy's main qualifications were his guns and his fists, but he was followed by men who could cope with the paperwork, had a background in either some field of biology or wildlife law enforcement, and were more willing to follow policy directives from those further up the hierarchy.

My mother's father, Dr. Robert S. Norton—"Doc" to everybody but his family—was one of these. He was born on a ranch in Oklahoma while it was still a territory. After earning his degree as a doctor of veterinary medicine in Chicago he set up a practice in Velva, North Dakota. There he met and married my grandmother, a teacher who had homesteaded nearby, and there their three children were born. Granddad served as a medical officer in World Wars I and II. At the end of World War I he joined the U.S. Fish and Wildlife service. He became superintendent of the National Bison Range in 1930. My father was working there and married his new boss's daughter. My parents and my older brother, Bob, were living in a house on the Range when I was born in 1933.

My grandfather, with his ranching background, was comfortable with the policy Andy had rejected. Since there was no market for live bison, excess animals were slaughtered and the meat sold. During Granddad's time a federal bison manager was judged largely by the number of bison he produced and maintained. Everybody thought the National Bison Range could support lots of bison. Professor Elrod had surveyed the site and confidently predicted the range could support 1,200 head plus lots of elk and deer. It couldn't. Populations of 600 to 700 head overgrazed the range. Granddad fed them hay in the winter, but even so the range was hammered; his photos show bison standing on nearly bare ground—their hooves completely visible.

Granddad left the National Bison Range shortly before World War II started and for twenty years his successors were senior wardens—arriving a few years short of retirement. Like most bison managers of their generation, they were competent with the paperwork and committed to protecting the place and the animals. But they didn't see the Range as an ecological community and weren't preoccupied with the condition of the grasses. In the early 1960s a new wave came in. When I returned to the National Bison Range in 1965—a newly minted Ph.D. intending to study the behavioral ecology of bison—Joe Mazonni had just taken charge. Young and educated in ecology, he saw the bison range as a community of animals based on a community of plants. Biologists first and bureaucrats second, this generation looked at the grassland beneath the buffalo's feet and shook their heads. With the grasses grazed nearly to their roots, they were losing their competition with unpalatable plants, many of them exotic. Those in Joe's generation knew that the first requirement of grazers is grass, and they set out to restore it.

And they were working not just to feed the bison but also to restore the grassland in its own right. The biggest problem was that bison use good judgment when they graze: they choose the tasty and nutritious grasses and leave the other plants. The bison's grazing choices created a sort of natural selection favoring inedible plants. However, if the native grasses go ungrazed for a whole growing season, they can gain ground even against many of the exotic plants that have invaded the North American grasslands during the last two centuries. So the National Bison Range, like other overgrazed bison refuges, was divided into several parts by fences. The bison are moved around to ensure that each of those parts goes ungrazed for a whole growing season every two or three years. Now the grasses can compete. In some places the range managers are trying a technique that the plains Indians used for centuries—setting the range on fire. One recent spring I parked beside a burned grassland on the Wichita Mountains Wildlife Refuge in Oklahoma. A short, vividly green crop of new grass carpeted the ground and the bison were drawn to its surface like iron filings to a magnet. It's slow work—thirty-five years hasn't been long enough to complete it—but the changes are in the right direction.

This generation, and those that followed, saw the Range as a community that should be nudged toward a natural state. One can only nudge so far with a range in which most of the featured animals are not native— bighorn sheep, mountain goats, and the bison themselves are not native to this part of Montana—but you could let them function more naturally. My grandfather had hired a trapper to eradicate the coyotes; Joe instituted a live-and-let-live policy. To the local farmers and ranchers, from whom the summer staff was drawn, the break with tradition grated a bit. Frosty Largent was my dad's age, and I'd known him since I was a boy. Before I started my research he showed me around the Range and we talked about the new policies. Joe had said that the coyotes had a place in the scheme of things and should no longer be shot on sight. "He even told us the rattlesnakes have a place here." Frosty smiled ruefully and shook his head gently. "It's awful hard for a man not to kill a snake."

My dad had a cattleman's attitude toward the new grazing policy. "Ain't that just like the government," he would say, "lettin' all that feed go to waste. If I had that range I'd have three times the stock on it." It's just as well refuge managers aren't chosen by local elections.

The other public herds also grew; they eventually stabilized or were reduced, as the range required, managed by the same generation of managers with the same goals. With sizable herds in the public's hands, the short-term future of wild bison seemed assured.

THE INCREDIBLE SHRINKING GENE POOL

But what of the bison's long-term future? In the long run a species adapts by tapping its ultimate resource—its gene pool. The gene pool contains not just the species' reality but also its future possibilities. It limits not just what challenges a species can handle today, but its range of possible adjustments to future changes. Every species, no matter how large its gene pool, eventually exhausts its possibilities and goes extinct.

Many genes are in only some individuals. Therefore, the more individuals, the bigger the gene pool. Several million bison would have a very large gene pool, but reduce that population to several hundred and it becomes very likely that a lot of genes have fallen by the wayside of the road to extinction. By now, many wild species have traveled that route. We'd like to know how great the bison's loss has been and how that will affect their future. That knowledge would help us conserve bison and also give us insight into the future faced by animals with similar pasts.

DNA technology has brought one of the answers tantalizingly close. Theoretically we can compare the DNA of buffalo past to that of buffalo present. DNA is in both blood and bones. We have bones of animals that died before the population crashed and the blood of today's bison. It should be a simple matter to compare the past DNA complement to the present. It's straightforward, but it isn't easy. I'm one of several who've tried. Our team consisted of me; Dirk van Vuren, who knows bison and museum collections; and Cecelia Torres-Penedo, a geneticist specializing in the cattle family and experienced with bison genetics.

Working with old bones is called *forensic analysis*—it incorporates all the concerns about specimen identification and sample contamination that an FBI lab must deal with. There was the added complication that unless the bones were several hundred years old they were of little use. We examined three museum-age samples—about 200 years old—and got some signals from two. But there were no signals the blood of current bison wasn't sending.

We cast about for more bones and George Frison, an anthropologist and an expert on prehistoric North American bison hunting, generously gave us three precious several-hundred-year-old samples from one of his digs. Cecelia couldn't raise a signal from any of the three. Others using the same approach have had similarly limited success. It's the way to go, but we're not far along it yet. And until we are a good bit further, we won't be able to compare the present bison's potential to evolve to that of bison 300 years ago.

In the meantime, the surviving population has been divided into even smaller populations in parks and refuges. The size of a species gene pool sets a limit on future adaptation. But very small populations raise another specter: inbreeding. Today's plains bison—*Bison bison*—mingled with millions of their own kind drifting across a wide and unbroken sea of grass. Suddenly they were reduced to scattered handfuls confined to tiny islands a few miles across that more or less matched their original habitat. It's a scenario sure to chill the blood of a genetically oriented conservation biologist. Deleterious genes that would have been diluted to near-insignificance in a gene pool contributed to and drawn from by millions could suddenly be concentrated and vigorously expressed in a gene pool drawn from fewer than a dozen animals.

From widely outbred to severely inbred in one or two generations—the worst possible case of the infamous genetic bottleneck. It all adds up to a gloomy forecast for the American bison. Just how gloomy it is and what we can possibly do to improve it were the central questions addressed by the conservation biologists Joel Berger and Carol Cunningham. A critical step in getting answers is to determine the size of the effective breeding population—the number of animals in a particular population that actually transmit genes to the next generation, and how many offspring each has. The smaller the effective breeding population, the greater the risk of inbreeding. If every adult had the same number of offspring you could get the number by simply counting. But they don't.

In fact, a large part of bison social behavior functions to enable some individuals to leave more offspring than others.

The breeding system is critical. Bison breed promiscuously and more dominant bulls father more calves. In a one-year study I saw some bulls breed five cows and others none. In a three-year study Jerry Wolff saw one bull breed sixteen cows and another none. Joel and Carol studied the Badlands National Park herd for four years and saw one bull breed twenty-eight cows and another breed none. They saw some cows produce five calves in five years and others produce none. My graduate student John Galland and I studied cows on Catalina Island, California, for four years and saw some cows produce four calves and others produce none. So there are big differences between individuals, and bigger differences between individual bulls than between individual cows.

During three years of their study, Joel and Carol could identify an average of 137 animals each year, of which 68 were breeding-age adults. When they calculated the effective breeding population of these animals over those years, taking into account individual differences in the reproductive success of both bulls and cows, the breeding population turned out to be considerably lower: as few as 21 by one of the computational models conservationists rely on, and not more than 46 by any model. That's well below the number needed to prevent inbreeding. Still, the more years the herd was studied the larger the calculated effective breeding population became, so a longer study would probably have produced a higher estimate.

Much of the disparity between the total number of adult bulls and cows and the effective breeding population is the consequence of great differences between bulls. An extreme example will make the point. Suppose that a population consisted of ten bulls and ten cows, and that one dominant bull bred all the cows. Half the genes in the next generation would come from one individual. The effective breeding population would not be twenty or eleven, but fewer than four.

Joel and Carol's analysis suggests that the population of about 400 bison in Badlands National Park will lose heterozygosity at the rate of 1 percent per generation. Given that at present only Yellowstone Park has a significantly larger population, an acceptable effective breeding population can be achieved only by relocating females from herd to herd, thus

managing the several federal herds as a single meta-population. Theoretically, it would be more efficient to relocate males, but practically it isn't
so. Joel and Carol kept track of several cows and bulls introduced to the
Badlands Park from a population in Colorado. The Colorado bulls were
so intimidated by the Badlands bulls that not one of them ever bred. In
contrast, the Colorado cows had about as many calves as the Badlands
cows.

Perhaps someday we'll create at least one more island large enough to
accommodate a population large enough to be viable over the long term.

A NEW CONSERVATION CHALLENGE: ERADICATION BY MODIFICATION

More than 90 percent of the bison in North America today are undergoing domestication. People are not only taming the occasional individual
to pull a cart or carry a saddle, they're deciding, directly and indirectly,
which will be the parents of the next generation. They're aiming to en-
hance certain traits and selecting to parent the next generation those individuals that display them most strongly. I'm reading *Bison World*, a magazine for bison ranchers, and find an advertisement for bison people can
get along with. And I see a bull with an unusually wide rump photographed from behind with a wide-angle lens for breeding stock with
more of the higher-priced cuts. They're also often selecting against traits
incompatible with management strategies developed for cattle. Bison so
disturbed by crowding that they succumb to stress disease, bison so resistant to close confinement that they injure themselves trying to escape
from corrals, effectively select themselves out.

That's a bit of a gut-wrencher for me—from birth through the first
grade I lived on a refuge with bison. I studied and admired wild bison for
decades. The thought of their being converted to humpbacked cattle gets
my viscera going.

But when I engage my brain I can see lots of good things about domesticating bison. Conservation biologists often argue that one reason to
preserve genetic diversity is that useful things may come from those
genes. Well, what could be more useful than new domestic livestock,

preadapted to our continent and easy on the soil that must support its weight and appetite? Livestock, moreover, whose flesh is not as fatty as European cattle's and is said to be less allergy-provoking.

Domestication has by and large been a blessing. If some wild wolves hadn't been domesticated, we wouldn't have dogs and our lives would be the poorer for it. Besides, if we're careful we can have both wild and domestic bison, just as we can have both wolves and dogs. But our history with wolves and dogs should alert us to the danger that domestic bison pose to wild bison. The biologist Juliet Clutton-Brock and her colleagues have put together an instructive story. In the 1930s arctic wolves were exposed to casually managed husky sled dogs. Arctic wolf skulls began to look more and more like husky skulls until dog management tightened up and the wolf population began to dilute its husky genes. Tens of thousands of years of selection had given wolf skulls their shape—any change would almost certainly hurt its bearer. Moreover, it was almost certainly not the only change, but rather only the most visible of many changes, almost all of them worse for wolves.

Like wolves, wild bison still exist, but they exist as little islands in a sea of increasingly domesticated relatives—not yet cattle, but no longer wild bison. Perhaps we should call these animals that started as buffalo but will end as a kind of cattle "buffattle." If bison domestication goes well, buffalo ranching will spread and wild buffalo will end up on islands in a sea of buffattle. We must not let them drown in that sea. It would be a terrible irony if we saved wild buffalo from the hide hunters' Sharps rifle, then lost the species to the breeders' bottom line. The most vivid threat today is eradication by modification.

Buffattle are still developing, and we can't say exactly what they will be like. But we know enough about the domestication of other species to make some good guesses. We invented domestication long before we invented gunpowder. For at least 10,000 years humans have changed wild plants and animals to better meet our goals. Wolves yielded dogs, whose protection, affection, and hard work have so enriched human lives. North African wild cats yielded domestic cats, who ungrudgingly share their homes with us, so long as we keep up the supply of food, water, and laps. And then there are horses, burros, sheep, goats, yaks, water buffalo, chickens, ducks, turkeys, and many more animals along with hundreds of plants.

The aurochs yielded cattle: today there are scores of cattle breeds selected for everything from high milk production to sleeping sickness immunity. But what about the aurochs itself? All we have left are some bones and some cave paintings. The wild aurochs is gone—a victim not of its weaknesses but of its strengths. It offered humans too much for its own good. We plucked the many flowers, but discarded the bush. We don't know how the species died. Perhaps the last beating wild aurochs heart was pierced by a stone spearpoint and it bled to death, but more likely the few survivors of the species began to mate with their domesticated descendants, and it was bred to death.

Bison domestication is like hide hunting, except that instead of stripping off the hide and discarding the meat, bison domestication will strip out the genes that make for good domestic bison and discard the genes that make wild bison wild. By now it's clear which of the aurochs' genes our ancestors stripped off to save and which they discarded. They kept the genes that produced meat and milk; in fact, they soon realized that high-producing parents had high-producing offspring and selectively bred for more meat and milk. So they became breeders, selecting for animals that grew faster, bred younger, and were easier to intimidate.

At the same time these breeders discarded wild behavior. Cattle look a good deal like the wild aurochs, but even though we have no direct information about aurochs behavior, we can be sure it was radically different. Wildness, competitiveness, and self-protectiveness are vital to an animal living on its own, but they're a big nuisance to a rancher. Even a mother's strong attachment to her calf can be a nuisance. My dad raised cattle, and when a cow rejoining her calf ran over me that fact was pounded home by her hooves.

Self-protective animals will keep humans at a distance, and you can't ranch animals without getting close to them at least once in a while. The magazine advertisement for breeding stock boasts they are bison that people can live with. Males selected to compete with a lot of other males may remain preoccupied with competition when there are few males, and leave some cows unbred. We may admire a stallion's spirit, but when it comes to cattle we want wimps. Sedentary wimps at that. Restless, mobile animals use up calories and time that could be spent making more meat and milk.

So the needs of the rancher and the nature of wild bison clash head on. The rancher's goal has to be to take the wild out of the bison he or she is domesticating. The conservationist's goal has to be to preserve the wild in the remaining wild lines. It's perfectly possible that nobody gave a damn when the wild aurochs was being bred to death, and it's also not clear what anyone who did care could have done about it. But we, who for sure do give a damn, can keep wild bison from breeding to death, their blood diluted in a sea of buffattle.

Sometimes when I talk about wild bison someone points out that all of today's plains bison descend from animals that spent at least some time enclosed in a fence. Therefore, some argue, all of today's bison are domestic and there are no wild bison to preserve. That claim reflects a misunderstanding of what domestication is. It's not being confined—if it were, every animal in most zoos would be domesticated. They're not. Even those that take food from our hands are tamed—habituated to humans—not domesticated. The essence of domestication is selective breeding: humans deciding which individuals will produce the next generation, and choosing them to produce a next generation that will better serve human goals. That's the core of buffalo ranching.

Bison ranching isn't happening just on private property. As bison have become valuable, budgetary thinking has begun to look to the public herds for revenue. Managers are pressured to maximize the number of bison for sale. From that perspective each bison is a productive unit—valued for its contribution to the bottom line. Some public herd managers have embraced that perspective enthusiastically, others have accepted it reluctantly, but several do what it requires—boosting the number of cows on the range. Instead of a natural sex ratio of a bit less than one bull per cow (bulls don't live as long as cows in nature), the sex ratio of these herds is altered, perhaps to a ratio of one bull to ten cows.

Such ratios are right if your goal is to produce the maximum number of calves each year from a given amount of range. But while the range will produce more calves, they are likely to be less wild. Natural selection will select animals better suited to their circumstances. A biased sex ratio is a very important circumstance and seems certain to shift the breeding strategy of both males and females. It's likely that when there are lots of cows, bulls that back away from a challenge and spend the time and energy saved finding unattended cows will tend to father more calves. When there are few bulls, cows that aren't coy and don't run about early in estrus inciting competition between bulls will be more sure of being bred each year.

Moreover, selling bison means handling them in corrals. Individuals that attack their handlers are unlikely to have another chance to breed. In these and other ways the animals are being domesticated. Natural selection works and artificial selection works even faster. That's why wild bison behave the way they do, and why domestic bison will behave differently.

We have to learn to know the difference between them and we have to keep them apart. As a practical matter we have to take action in the opposite order: first separate them, then learn to know the difference. The separation is critical. As I am writing these words, the governor of the state of Maine is objecting to protection for wild Atlantic salmon in Maine. He argues that they have interbred with domestic Atlantic salmon escapees from fish farms, so there are no wild Atlantic salmon to protect. Maine's population, he says, has been bred to death.

How to recognize wild buffalo? We have to assume that the most wild are those whose life has been changed least by humans. They live on free ranges, they breed in herds composed of about as many males as females, and they find their own food summer and winter. If they have been tested by predators, so much the better. Nearly all the bison in North America today, wild and domestic, have recent forebears that lived like, and even with, cattle on turn-of-the-century ranches. Though human manipulation, even of the cattle, was minimal on these ranches, if we could have looked closely enough then we would likely have seen some evidence, however slight, of artificial selection acting on bison. But it's reasonable to expect that wilder circumstances have restored most if not all the wildness to today's wild bison.

Domestic bison will look very much like wild bison. That's why domesticating lines are such a subtle threat to wild lines. But their behavior will change profoundly and rapidly. It will behoove both conservationists and ranchers to recognize these behavioral changes. We'll need measures of behavior that are as reliable as the measures of the skulls of wolves hybridized with dogs but that can be taken on the living animal, so that when we select breeders for domestic lines, or crop wild herds to protect their range, we can make the right choices.

A better bison, from a stockman's point of view, would be less feisty and less restless. Feistiness takes energy and produces injuries—even deaths. Restlessness takes energy. We have a pretty good handle on the physiology of those traits and how domestication changes that physiology. Ed Price, a specialist in animal domestication, has pieced the story together. The underlying actors are neurotransmitters—chemicals released from one neuron and designed to act on other neurons. Two domestication neurotransmitters are serotonin and dopamine. The more serotonin in an animal, the less feisty it is. The less dopamine, the less active. Breeding for less feisty, less active animals works by artificially selecting for animals with more serotonin and less dopamine.

Low-dopamine animals will spend more time standing quietly and will walk, run, and play less. That will be easy to measure and quantify. High-serotonin bison will threaten one another less, and resist confinement in corrals and squeeze chutes less. A probable big quantifiable change will be in the distance at which they react to one another. The tricky thing is that measuring reaction requires another actor. The other actor might be hard to standardize, making reaction to it hard to quantify. In practice that means a great deal of variance, so the data will have to be analyzed statistically. Animal behaviorists have often used a mirror image to test for just this kind of difference between individuals. Very few species recognize their image as themselves. Most treat it as another individual of the same species and sex—a possible competitor or companion. Kenneth Armitage has used mirrors to quantify the personality traits of marmots. That would be a good way to go with bison, though bull proofing the mirror could be a bit challenging.

Ranching is a practical business—a bottom-line business. However much private owners—corporations, individual ranchers, or managers of tribal herds on reservations—may admire wild buffalo, economics precludes raising them. To keep bison wild you must be willing to lose money on them—or at least leave money on the table. Only the public can afford to do that, and only as publicly owned animals do wild bison have a future.

We need to find ways to make that future better than the present. The buffalo's hairbreadth escape from extinction was a conservation triumph and we are right to be proud of it. But we still haven't done very well by them. The number living as wild bison fluctuates dramatically as the populations in Yellowstone and Wood Buffalo Parks rise and fall, but there are typically fewer than 10,000 living as wild animals in parks and refuges. Most of them suffer from diseases introduced from Europe, and most live in populations so small that they will suffer gene loss.

The public is more than willing to lose money raising wild bison, yet most of the few places devoted to wild bison are small. Americans want more bison in more places. If we're going to keep wild bison wild, we're going to have to protect the spaces they have, and we should look for some more space—space on the grasslands that shaped them and that they shaped in turn. A grassland park, or parks, would help a lot. And we should be willing to consider resolving this paradox: *Bison bison* is the only wild animal in the United States that is not allowed to live as a wild animal—live outside parks and refuges—anywhere in its original range. Some people in Montana are vigorously advocating that wild bison should once again be a wildlife species.

Bison conservation evolves and changes. The problems—poaching, overgrazing, genetic dilution—have changed. The solutions—Andy's guns, easing the pressure on the grasses by grazing pastures in rotation, behavioral testing—have changed, or will change, to address them. What must never change is our commitment to the remarkable creatures that dominated the North American plains—shaking the earth when they stampeded and shaping the grasslands where they walked.

CHAPTER 25 A Great Plains Park

George Catlin, who traveled, wrote about, and painted the plains between 1832 and 1839, proposed a Great Plains Park created by the national government, where herds of elks and buffalo would be protected in perpetuity. Catlin was writing more than thirty years before Yellowstone became the world's first national park. And he was extolling the beauty of the Great Plains' biological community, not the spectacle of geysers, boiling mud, and rivers running in dramatic canyons they had worn through thousands of feet of rocks. Even today most of our parks and national monuments celebrate geological rather than biological phenomena. Catlin was way ahead of his time.

Imagine this—dawn on the Great Plains in late spring—just the first sliver of sun showing above the eastern horizon. Seeing the still steamy breath of a buffalo herd on a brisk morning. The cows graze to make milk to feed their calves, but pause to scan the horizon for wolves, bears, perhaps even mountain lions that could appear at any minute. Buffalo birds cluster round their feet, getting help feeding themselves and preparing to get help feeding their young.

The bison's hooves contrast with the bright green first flush of spring growth of the blue grama and buffalo grass and with the white or pink flowers of the prairie star and the dainty little bells of the yellow bell. The flower colors are calling in pollinators and beginning the cycle of flowers to seeds to next year's forbs that makes that prairie a pasture rather than a desert for the light-footed pronghorn daintily choosing them from the sea of grass.

In a prairie dog town, adults and young ease out of their burrows. They check the sky where hawks and eagles fly, repair mounds wallowing buffalo have deformed, and check the area for the ferrets, badgers, coyotes, and snakes that pass through town from time to time. The mothers are keeping killer kinfolk away from their nursery burrows. Burrowing owls are settling into their borrowed burrows for the day, having hunted through the night.

The bison exchange contact grunts. Distant pronghorn bucks challenge others with their twangy snort. It's too early in the year to hear many bison bellows or any elk bugles—these will come later—but the prairie dogs bark and jump-yip; the meadow larks and sparrows sing.

This was America's Serengeti—the scene that drew European wealth and royalty to the Great Plains as tourists 150 years ago. The infinite reach of the Great Plains vistas, the abundant animals—it was another world entered, profoundly different and dramatically beautiful. It was the scene Buffalo Bill Cody showed the Czarevitch.

The first American parks preserved oddities—generally beautiful ones: Yellowstone's geysers and hot mud, Yosemite's glacier-carved granite, the spectacular colors and shapes that water had revealed and sculpted in the Grand Canyon, Zion, Brice, and Arches. But they can also fill another role: to present a plausible reconstruction of primitive America. To preserve—or, if necessary, restore—living, functioning representations of the plant and animal communities that filled our land before European settlers came.

A Great Plains Park must be very large—at least 5,000 square miles—and must include both upland and river bottom habitat. It's too late to preserve such a representation of the Great Plains, but it's not too late to restore one. Canada has shown the way and even pointed to a place. Its story begins in 1956, when the Saskatchewan Natural History Society began to push for a Grasslands National Park. Fast-forward to 1981, when the federal government and Saskatchewan formally agreed to establish Grasslands National Park in southwestern Saskatchewan. The area selected was largely uncultivated, privately owned grassland acquired from willing sellers by a willing buyer—Parks Canada. The completed park will be in two blocks totaling 350 square miles. As of 1999 the east block was 49 percent acquired; the west block, 61 percent. Now Parks Canada is dealing with the devilish details of maintaining and restoring the biological integrity of the land. On the uncultivated rangeland cattle replaced bison as the primary grazers. As the cattle leave they must be replaced, and bison are a logical replacement.

The park's location was chosen primarily because the resources to accomplish its mission are there. But it is also an undisguised invitation to the United States. Both blocks lie on eastern Montana's northern border along a political dividing line through a biological unit. The same grassland lies south of the border, Parks Canada points out, and it goes on to note that the U.S. federal government owns 31 percent of that grassland region. So Canada has given us a big hint, and seems ready to welcome a U.S.-Canada grassland with an open border.

And there are many other possible places. The federal government already owns large short-grass and mixed-grass tracts, and many more of both are for sale. So a grassland park in the United States is possible—and would not even be difficult to accomplish. It hardly needs saying that such a park would give a welcome boost to wild bison conservation. It's too early to say exactly where, but we and our prairie heritage deserve, need, and surely will someday have at least one Great Plains National Park. Catlin called for one in the first part of the nineteenth century. Perhaps we can have one early in the twenty-first.

Notes

CHAPTER 1. BULL TO BULL AND COW TO BULL

pp. 5–13. On aggression and communication in bulls, see McHugh 1972; Lott 1974a, 1974b.

pp. 13–14. On bison getting bolder as they get older, see Maher and Byers 1987; see also Komers, Messier, and Gates 1994.

pp. 14–19. On bulls tending cows, see McHugh 1972; Lott 1974b, 1976, 1979a, 1981.

p. 17. On breeding and birth synchrony in moose and wildebeest, see Whittle et al. 2000; Estes and Estes 1999.

pp. 16–18. On the behavior of pre-estrous cows, see Lott 1976, 1981; Komers, Messier, Flood, and Gates 1994.

p. 16. On cows' choice of mate, see Wolff 1998.

pp. 18–19. On bison copulation, see McHugh 1972; Lott 1976, 1981.

p. 20. On bulls sequestering cows, see Lott 1981.

p. 20. On the breeding advantage of dominant bulls, see Lott 1979a; Wolff 1998; Berger and Cunningham 1994.

p. 20. On bulls' loss of hair, see Lott 1979b.

CHAPTER 2. COW TO COW

p. 23. On aggression rates among cows, see Rutberg 1986.

p. 24. On the benefits of dominance, see Rutberg 1986.

pp. 25–27. On the determinants of dominance, see Rutberg 1983; Lott and Galland 1987.

CHAPTER 3. COW TO CALF

pp. 29–30. For a description of bison giving birth, see Lott and Galland 1985b.

p. 30. On the hider strategy of pronghorn, see Byers 1997.

p. 31. On the follower strategy of bison, see Green and Rothstein 1993.

p. 32. On the end of cow-calf attachment, see Lott and Minta 1983b; Green 1993.

p. 33. On the sites where bison give birth, see Lott and Galland 1985b.

p. 34. On parent-offspring conflict, see Green, Rothstein, and Griswold 1993.

CHAPTER 4. BISON ATHLETICS

p. 42. On bison running and jumping, see Guthrie 1990.

pp. 43–45. On animal locomotion in general, see Hildebrand et al. 1985.

p. 45. On bison fighting, see Guthrie 1990; Geist 1996.

CHAPTER 5. DIGESTION

pp. 48–50. On ruminant digestion, see van Soest 1994.

CHAPTER 6. TEMPERATURE CONTROL

p. 55. On the density of bison's hair coat, see Mooring and Samuel 1998.

CHAPTER 7. ANCESTORS AND RELATIVES

p. 61. On Blue Babe, see Guthrie 1990.

pp. 62–64. On bison evolution in North America, see Guthrie 1990; McDonald 1981; Wyckoff and Dalquest 1997.

p. 64. On the Pleistocene overkill theory, see Martin and Klein 1984; on the effect of human hunters, see Guthrie 1990.

p. 67. On wood bison trails, see Carbyn, Oosenbrug, and Anions 1993.

p. 67. On the distinguishing features of wood bison, see Geist and Karsten 1977; van Zyll de Jong et al. 1995.

p. 68. On the DNA results comparing wood and plains bison, see Wilson and Strobeck 1999.

p. 68. For the argument that wood and plains bison are merely ecotypes, see Geist 1991.

CHAPTER 8. HOW MANY?

p. 70. On Shaw's tracing of the figure of 60 million bison and other estimates of herd size, see Shaw 1995.

p. 70. On Dodge's wagon ride, see Dodge [1877] 1959.

p. 70. For Seton's two calculations, see Seton 1910; 1927, vol. 3. For Hornaday's estimate, see Hornaday [1887] 1971, 391.

p. 72. For bison numbers west of the Rockies, see van Vuren 1987.

p. 73. For Flores's reanalysis of USDA data, see Flores 1991.

p. 73. For McHugh's approach to figuring carrying capacity, see McHugh 1972.

p. 74. For Haynes's estimate of the carrying capacity of part of Montana, see Haynes 1998.

p. 75. On the effects of wolves on bison, see Carbyn, Oosenbrug, and Anions 1993.

p. 75. On the killing winter of 1841, see Dodge [1877] 1959.

CHAPTER 9. THE CENTRAL GRASSLAND

p. 81. On the "sea of grass," see Manning 1995; L. Brown 1985; Hunt 1974.

p. 81. On the soils of the American Prairie, see Hunt 1974.

p. 82. On the ecologists' division of the American Prairie, see Brown 1985; Manning 1995; Knopf and Samson 1997.

pp. 82–84. On the various prairie grasses, see Brown 1985; Manning 1995; Knopf and Samson 1997.

p. 85. On fire and the movement of bison, see Shaw and Carter 1990.

p. 87. On the benefits of bison urine, see Knopf and Samson 1997.

p. 88. On wallows, see Knopf and Samson 1997.

p. 89. For Schaller on the Serengeti, see Schaller 1972, 10.

pp. 89–91. On the history of the grassland, see Stegner 1954; Manning 1995.

p. 92. On the Poppers' proposal, see Matthews 1992.

CHAPTER 10. WOLVES AND BISON

p. 99. On wolves killing calves, see Carbyn, Oosenbrug, and Anions 1993.

p. 100. On Pleistocene animals, see Kurtén 1988; Guthrie 1990.

pp. 101–102. On wolf society, see Mech 1991.

p. 102. For Moehlman's observations on dog litters, see Moehlman 1996.

p. 103. On the usual location of calves within a herd, see Carbyn, Oosenbrug, and Anions 1993.

p. 103. On the marathon standoff between wolves and bison, see Carbyn and Trottier 1988.

CHAPTER 11. BUFFALO BIRDS

p. 105. On female buffalo birds, see Davies 2000.

p. 105. On birds that are brood parasites, see Davies 2000.

p. 106. On the modification of song in response to the reactions of female and male buffalo birds, see M. West, King, and Freeburg 1997; Davies 2000.

CHAPTER 12. DISEASES AND PARASITES

p. 108. On bison's single-celled parasites, see Alcamo 1994.

p. 109. On the transmission of *B. abortus,* see Cheville, McCullough, and Paulson 1998; Baskin 1998.

p. 114. On anthrax, see Alcamo 1994.

p. 115. On the ticks' quest, see Sonenshine 1991–93.

p. 116. On winter ticks and bison, see Mooring and Samuel 1998.

p. 117. On ticks and bison calves, see Mooring and Samuel 1998.

p. 117. On the grooming clock of African antelopes, see B. Hart 1997; B. Hart et al. 1992; L. Hart, Hart, and Wilson 1996.

p. 118. On modulation of the grooming rate, see B. Hart 1997.

pp. 118–19. On ticks' breeding strategies, see Sonenshine 1991–93.

CHAPTER 13. PRONGHORN

p. 122. For Kitchen's research, see Kitchen 1974.

p. 123. On the dwindling of pronghorn territoriality, see Byers and Kitchen 1988.

p. 125. On the quantification of territoriality, see Maher 2000.

p. 126. On the territoriality of white rhinos, see Owen-Smith 1988.

p. 126. On horses becoming territorial, see Rubenstein 1981.

CHAPTER 14. PRAIRIE DOGS

p. 127. On the numbers of prairie dogs in the nineteenth century, see Hoogland 1995.

p. 127. On the design of prairie dog burrows, see Hoogland 1995.

p. 128. On the ways that bison benefit prairie dogs, see Knopf and Samson 1997.

pp. 129–30. On the killing of pups by female relatives, see Hoogland 1995.

p. 131. On ground squirrels' response to large and small rattlesnakes, see Owings and Loughry 1985; Owings and Morton 1998; Swaisgood, Rowe, and Owings 1999.

p. 132. On burrowing owls imitating snakes, see Rowe, Coss, and Owings 1986; Owings and Morton 1998.

p. 132. On baby owls adding themselves to neighboring broods, see Johnson 1993a.

CHAPTER 15. BADGERS

p. 134. On the social organization of badgers, see Minta 1993; Goodrich and Buskirk 1998.

p. 134. On the economics of the territoriality of nectar-eating birds, see Gill and Wolf 1975; Pyke 1979; Lott and Lott 1991.

pp. 136–37. On coyote-badger partnerships, see Minta, Minta, and Lott 1992.

CHAPTER 16. COYOTES

p. 138. On the range of coyotes, see Bekoff 1978.

p. 138. On the behavior of coyotes in arid regions, see Andelt 1985.

p. 139. On coyotes killing grown mule deer, see Bowen 1981.

p. 139. On family packs of coyotes, see Bekoff 1978.

p. 140. On poisoning coyotes for pelts, see Robinson 1995.

CHAPTER 17. GRIZZLIES

p. 142. On the scavenging of hungry bears, see Mattson 1997.

p. 143. On sexually selected infanticide in various species, see Hausfater and Hrdy 1984.

CHAPTER 18. FERRETS

pp. 145–46. On black-footed ferrets, see Miller and Anderson 1993.

CHAPTER 19. CLOSE ENCOUNTERS

p. 151. On feeding bighorn sheep, see Lott 1988.

p. 153. For the story of Dick Clark and his bull, see Dary 1974.

p. 155. On the killing of A. H. Cole by his buffalo bulls, see Dary 1974.

p. 157. For the recommendation to stand your ground when a buffalo charges, see McHugh 1972.

CHAPTER 20. TO KILL A BISON

pp. 159–60. On buffalo jumps, see Frison 1998.

p. 164. On the technique of stand hunters, see Robinson 1995; Isenberg 2000.

p. 164. On Mooar, see Robinson 1995.

CHAPTER 21. NUMBERS BEFORE THE GREAT SLAUGHTER

p. 167. For "The Dwindle," see Seton 1927.

p. 168. For Hornaday's estimate, see Hornaday [1887] 1971; quotation, 391.

p. 168. "One may assume . . . ": Shaw 1995, 150.

CHAPTER 22. WHERE HAVE ALL THE BISON GONE?

p. 170. For the standard story of the disappearance of the buffalo, see McHugh 1972; Dary 1974.

p. 170. On the buffalo's vanishing at different times for different reasons, see Krech 1999; Isenberg 2000.

p. 171. On the Spanish trade in buffalo robes and hides, see Creel 1991.

p. 171. On the effect of horses on buffalo hunting, see Flores 1991; Isenberg 2000.

p. 172. On famine among the Comanches, see Flores 1991; E. West 1995; Isenberg 2000.

p. 172. The figure of nine or ten bulls for every cow is found in Morgan 1998. For Parkman's account, see Parkman 1849, 401.

p. 173. On the role of the Métis in supplying the most robes, see Foster 1992.

pp. 174–75. For Ross's account of a Red River hunt, see MacEwan 1995.

p. 175. On the trade in goods once considered luxuries, see Foster 1992.

p. 175. On the special excursion trains for hunting buffalo, see McHugh 1972; Dary 1974.

p. 176. On the use of buffalo leather in industrial belting, see Robinson 1995; Isenberg 2000.

p. 176. On Mooar's first sale, see Robinson 1995.

p. 176. On the story about General Sheridan, see McHugh 1972; Robinson 1995.

p. 178. On Harris's buffalo hunt in Texas (mentioned in chapter 16), see Robinson 1995.

p. 179. For Huffman's photos, see M. Brown and Felton 1956.

CHAPTER 23. ATTITUDES

p. 180. The engraving appears in *Harper's Weekly,* December 12, 1874.

p. 181. On the new social context of the hide hunt, see Isenberg 2000.

p. 182. "Except an elephant . . . ": Parkman 1849, 401.

p. 182. "I did not want to shoot . . . ": Orwell 1950, 7.

p. 183. On Mayer, see Mayer and Roth 1958.

CHAPTER 24. CONSERVATION

pp. 185–86. For the story of Samuel Walking Coyote, see Dary 1974.

p. 193. On the bison's shrinking gene pool, see Berger and Cunningham 1994.

p. 194. On determining the size of an effective breeding population, see Berger and Cunningham 1994.

p. 194. For Wolff's three-year study, see Wolff 1998.

p. 194. For the data from the Badlands, see Berger and Cunningham 1994.

p. 194. For data from Catalina Island, see Lott and Galland 1985a.

p. 196. On arctic wolves, see Clutton-Brock, Kitchner, and Lynch 1994.

p. 200. For an analysis of how domestication changes physiology, see Price 1999.

p. 200. On the personality traits of marmots, see Armitage 1986.

CHAPTER 25. A GREAT PLAINS PARK

p. 202. For Catlin on the West, see Catlin 1989.

Bibliography

Alcamo, I. E. 1994. *Fundamentals of Microbiology.* 4th ed. Redwood City, Calif.: Benjamin/Cummings.

Andelt, W. F. 1985. "Behavioral Ecology of Coyotes in South Texas." *Wildlife Monographs* 94: 1–45.

Armitage, K. B. 1986. "Individuality, Social Behavior, and Reproductive Success in Yellow-Bellied Marmots (*Marmota flaviventra*)." *Ecology* 67: 1186–93.

Barsness, L. 1977. *The Bison in Art: A Graphic Account of the American Bison.* Flagstaff, Ariz.: Northland Press.

Baskin, Y. 1998. "Home on the Range." *Bioscience* 48: 245–51.

Bekoff, M. E. 1978. *Coyotes: Biology, Behavior, and Management.* New York: Academic Press.

Berger, J., and C. Cunningham. 1994. *Bison: Mating and Conservation in Small Populations.* New York: Columbia University Press.

Bowen, W. D. 1981. "Variation in Coyote Social Organization: The Influence of Prey Size." *Canadian Journal of Zoology* 59: 639–52.

Brown, L. 1985. *Grasslands.* New York: Alfred K. Knopf.

Brown, M. H., and W. R. Felton. 1956. *Before Barbed Wire: L. A. Huffman, Photographer on Horseback.* New York: Holt.

Buehler, K J. 1997. "Where's the Cliff? Late Archaic Bison Kills in the Southern Plains." *Plains Anthropologist* 42: 135–43.

Byers, J. A. 1997. *American Pronghorn: Social Adaptations and the Ghosts of Predators Past.* Chicago: University of Chicago Press.

Byers, J. A., and D. W. Kitchen. 1988. "Mating System Shift in a Pronghorn Population." *Behavioral Ecology and Sociobiology* 22: 355–60.

Callenbach, E. 1996. *Bring Back the Buffalo! A Sustainable Future for America's Great Plains.* Washington, D.C.: Island Press.

Carbyn, L. N., N. J. Lunn, and K. Timoney. 1998. "Trends in the Distribution and Abundance of Bison in Wood Buffalo National Park." *Wildlife Society Bulletin* 26: 463–70.

Carbyn, L. N., S. M. Oosenbrug, and D. W. Anions. 1993. *Wolves, Bison, and the Dynamics Related to the Peace-Athabasca Delta in Canada's Wood Buffalo National Park.* [Edmonton]: Canadian Circumpolar Institute, University of Alberta.

213

Carbyn, L. N., and T. Trottier. 1987. "Responses of Bison on Their Calving Grounds to Predation by Wolves in Wood Buffalo National Park." *Canadian Journal of Zoology* 65: 2072–78.

———. 1988. "Descriptions of Wolf Attacks on Bison Calves in Wood Buffalo National Park." *Arctic* 41: 297–302.

Catlin, G. 1989. *North American Indians*. Edited by P. Matthiessen. New York: Penguin.

Cheville, N. F., and D. R. McCullough. 1998. *Brucellosis in the Greater Yellowstone Area*. Washington, D.C.: National Academy Press.

Clutton-Brock, J., A. C. Kitchner, and J. M. Lynch. 1994. "Changes in the Skull Morphology of the Arctic Wolf, *Canis lupus arctos,* during the Twentieth Century." *Journal of Zoology* 233: 19–36.

Collins, S. L., A. K. Knapp, J. M. Briggs, J. M. Blair, and E. M. Steinauer. 1998. "Modulation of Diversity by Grazing and Mowing in Native Tallgrass Prairie." *Science* 280: 745–47.

Coppedge, B. R., and J. H. Shaw. 1997. "Effects of Horning and Rubbing Behavior by Bison (*Bison bison*) on Woody Vegetation in a Tallgrass Prairie Landscape." *American Midland Naturalist* 138: 189–96.

Creel, D. 1991. "Bison Hides in Late Prehistoric Exchange in the Southern Plains." *American Antiquity* 56 (1): 40–49.

Dary, D. A. 1974. *The Buffalo Book: The Full Saga of the American Animal*. Chicago: Sage Books.

Davies, N. B. 2000. *Cuckoos, Cowbirds, and Other Cheats*. London: T. and A. D. Poyser.

Desmond, M. J., J. A. Savidge, and T. F. Seibert. 1995. "Spatial Patterns of Burrowing Owl (*Speotyto cunicularia*) Nests within Black-Tailed Prairie Dog (*Cynomys ludovicianus*) Towns." *Canadian Journal of Animal Science* 73: 1375–79.

Dodge, R. I. [1877] 1959. *The Plains of the Great West and Their Inhabitants; Being a Description of the Great Plains, Game, Indians, &c. of the Great North American Desert*. Reprint. New York: Archer House.

Dragon, D. C., B. T. Elkin, J. S. Nishi, and T. R. Ellsworth. 1999. "A Review of Anthrax in Canada and Implications for Research on the Disease in Northern Bison." *Journal of Applied Microbiology* 87: 208–13.

Estes, R. D., and R. K. Estes. 1979. "The Birth and Survival of Wildebeest Calves." *Zeitschrift für Tierpsychologie* 50: 45–95.

Flores, D. 1991. "Bison Ecology and Bison Diplomacy: The Southern Plains from 1800 to 1850." *Journal of American History* 78: 465–85.

Foster, J. E. 1992. "The Métis and the End of the Plains Buffalo in Alberta." In *Buffalo,* edited by J. E. Foster, D. Harrison, and I. S. MacLaren, 61–77. Edmonton: University of Alberta Press.

Foster, J. E., D. Harrison, and I. S. MacLaren, eds. 1992. *Buffalo.* Edmonton: University of Alberta Press.

Frison, G. C. 1998. "Paleoindian Large Mammal Hunters on the Plains of North America." *Proceedings of the National Academy of Sciences of the United States of America* 95: 14576–83.

Galbraith, J. K., G. W. Mathison, R. J. Hudson, T. A. McAllister, and K. J. Cheng. 1998. "Intake, Digestibility, Methane, and Heat Production in Bison, Wapiti, and White-Tailed Deer." *Canadian Journal of Animal Science* 78: 681–91.

Gates, C. C., B. T. Elkin, and D. C. Dragon. 1995. "Investigation, Control, and Epizootiology of Anthrax in a Geographically Isolated, Free-Roaming Bison Population in Northern Canada." *Canadian Journal of Veterinary Research* 59: 256–64.

Geist, V. 1991. "Phantom Subspecies: The Wood Bison *Bison bison 'athabascae' Rhoads* 1897 Is Not a Valid Taxon, but an Ecotype." *Arctic* 44: 283–300.

———. 1996. *Buffalo Nation: History and Legend of the North American Bison.* Stillwater, Minn.: Voyageur Press.

Geist, V., and P. Karsten. 1977. "The Wood Bison (*Bison bison athabascae Rhoads*) in Relation to Hypotheses on the Origin of the American Bison (*Bison bison Linnaeus*)." *Zeitschrift für Saugetierkunde* 42: 119–27.

Gill, F. B., and L. L. Wolf. 1975. "Economics of Feeding Territoriality in the Golden-Winged Sunbird." *Ecology* 56: 333–45.

Goodrich, J. M., and S. W. Buskirk. 1998. "Spacing and Ecology of North American Badgers (*Taxidea taxus*) in a Prairie-Dog (*Cynomys leucurus*) Complex." *Journal of Mammalogy* 79: 171–79.

Green, W. C. H. 1990. "Reproductive Effort and Associated Costs in Bison (*Bison bison*): Do Older Mothers Try Harder?" *Behavioral Ecology* 1: 148–60.

———. 1992. "The Development of Independence in Bison: Pre-weaning Spatial Relations between Mothers and Calves." *Animal Behaviour* 43: 759–73.

———. 1993. "Social Effects of Maternal Age and Experience in Bison: Pre- and Post-weaning Contact Maintenance with Daughters." *Ethology* 93: 146–60.

Green, W. C. H., and A. Rothstein. 1993. "Asynchronous Parturition in Bison: Implications for the Hider-Follower Dichotomy." *Journal of Mammalogy* 74: 920–25.

Green, W. C. H., A. Rothstein, and J. G. Griswold. 1993. "Weaning and Parent-Offspring Conflict: Variation Relative to Interbirth Interval in Bison." *Ethology* 95: 105–25.

Guthrie, R. D. 1990. *Frozen Fauna of the Mammoth Steppe: The Story of Blue Babe.* Chicago: University of Chicago Press.

Halpin, Z. T. 1983. "Naturally Occurring Encounters between Black-Tailed Prairie Dogs (*Cynomys ludovicianus*)." *American Midland Naturalist* 109: 50–54.

Hart, B. L. 1997. "Behavioural Defence." In *Host-Parasite Evolution: General Principles and Avian Models,* edited by D. H. Clayton and J. Moore, 59–77. Oxford: Oxford University Press.

Hart, B. L., L. A. Hart, M. S. Mooring, and R. Olubayo. 1992. "Biological Basis of Grooming Behavior in Antelope: The Body-Size Vigilance and Habitat Principles." *Behaviour* 44: 615–31.

Hart, L. A., B. L. Hart, and V. J. Wilson. 1996. "Grooming Rates in Klipspringer and Steinbok Reflect Environmental Exposure to Ticks." *African Journal of Ecology* 34: 79–82.

Hausfater, G., and S. B. Hrdy, eds. 1984. *Infanticide: Comparative and Evolutionary Perspectives.* New York: Aldine.

Haynes, T. 1998. "Bison Hunting in the Yellowstone River Drainage, 1800–1884." In *International Symposium on Bison Ecology and Management in North America,* edited by L. Irby and J. Knight, 303–11. Bozeman: Montana State University.

Hildebrand, M., et al., eds. 1985. *Functional Vertebrate Morphology.* Cambridge, Mass.: Harvard University Press, Belknap Press.

Hoogland, J. L. 1995. *The Black-Tailed Prairie Dog: Social Life of a Burrowing Mammal.* Chicago: University of Chicago Press.

Hornaday, W. T. [1887] 1971. *The Extermination of the Bison.* Facsimile reprint. Seattle: Shorey Book Store.

Huebner, J. A. 1991. "Late Prehistoric Bison Populations in Central and Southern Texas." *Plains Anthropologist* 36: 343–58.

Hunt, C. B. 1974. *Natural Regions of the United States and Canada.* San Francisco: W. H. Freeman.

Isenberg, A. C. 2000. *The Destruction of the Bison: An Environmental History, 1750–1920.* Cambridge: Cambridge University Press.

Johnson, B. S. 1993a. "Characterization of Population and Family Genetics of the Burrowing Owl by DNA Fingerprinting." *Journal of Raptor Research* 27: 89.

———. 1993b. "Reproductive Success, Relatedness, and Mating Patterns in a Colonial Bird, the Burrowing Owl." *Journal of Raptor Research* 27: 61.

Kitchen, D. W. 1974. *Social Behavior and Ecology of the Pronghorn.* Wildlife Monographs 38. [Ithaca, N.Y.]: Wildlife Society.

Knopf, F. L., and F. B. Samson, eds. 1997. *Ecology and Conservation of Great Plains Vertebrates.* New York: Springer.

Komers, P. E., F. Messier, P. F. Flood, and C. C. Gates. 1994. "Reproductive Behavior of Male Wood Bison in Relation to Progesterone Level in Females." *Journal of Mammalogy* 75: 757–65.

Komers, P. E., F. Messier, and C. C. Gates. 1994. "Plasticity of Reproductive Behavior in Wood Bison Bulls: On Risks and Opportunities." *Ethology, Ecology, and Evolution* 6: 481–95.

Krech, S. I. 1999. *The Ecological Indian*. New York: Norton.

Kurtén, B. 1988. *Before the Indians*. New York: Columbia University Press.

Lott, D. F. 1972a. "Bison Would Rather Breed Than Fight." *Natural History* 81 (7): 40–46.

———. 1972b. "The Way of the Bison: Fighting to Dominate." In *The Marvels of Animal Behavior,* [edited by T. B. Allen], 321–32. [Washington, D.C.]: National Geographic Society.

———. 1974a. *Aggressive Behavior in Mature Male American Bison*. 16mm, color, 12 min. University Park: Pennsylvania State University, Audio Visual Services.

———. 1974b. "Sexual and Aggressive Behavior of Adult Male American Bison (*Bison bison*)." In *The Behaviour of Ungulates and Its Relation to Management,* edited by V. Geist and F. Walther, 1:382–94. Morges, Switzerland: International Union for the Conservation of Nature and Natural Resources.

———. 1976. *Sexual Behavior in the American Bison*. 16mm, color, 9 min. University Park: Pennsylvania State University, Audio Visual Services.

———. 1979a. "Dominance Relations and Breeding Rate in Mature Male American Bison." *Zeitschrift für Tierpsychologie* 49: 418–32.

———. 1979b. "Hair Display Loss in Mature Male American Bison: A Temperate Zone Adaptation?" *Zeitschrift für Tierpsychologie* 49: 71–76.

———. 1981. "Sexual Behavior and Intersexual Strategies in American Bison." *Zeitschrift für Tierpsychologie* 56: 97–114.

———. 1983. "The Buller Syndrome in American Bison Bulls." *Applied Animal Ethology* 11: 183–86.

———. 1988. "Feeding Wild Animals: The Urge, the Interaction, and the Consequences." *Anthrozoos* 1: 255–57.

———. 1991. "American Bison Socioecology." *Applied Animal Behaviour Science* 29: 135–45.

———. 1993. "Lens Length Predicts Mountain Goat Disturbance." *Anthrozoos* 5: 254–55.

———. 1998. "Impact of Domestication on Bison Behavior." In *International Symposium on Bison Ecology and Management in North America,* edited by L. Irby and J. Knight, 103–6. Bozeman: Montana State University.

Lott, D. F., K. Benirschke, J. N. McDonald, C. Stormont, and T. Nett. 1993. "Physical and Behavioral Findings in a Pseudohermaphrodite American Bison." *Journal of Wildlife Diseases* 29: 360–63.

Lott, D. F., and J. C. Galland. 1985a. "Individual Variation in Fecundity in an American Bison Population." *Mammalia* 49: 300–302.

———. 1985b. "Parturition in American Bison: Precocity and Systematic Variation in Cow Isolation." *Zeitschrift für Tierpsychologie* 69: 66–71.

———. 1987. "Body Mass as a Factor Influencing Dominance Status in American Bison Cows." *Journal of Mammalogy* 68: 683–85.

Lott, D. F., and D. Y. Lott. 1991. "Bronzy Sunbirds (*Nectarinia kilimensis*) Relax Territoriality in Response to Internal Changes." *Ornis Scandinavica* 22: 303–7.

Lott, D. F., and M. McCoy. 1995. "Asian Rhinos on the Run? Impact of Tourist Visits on One Population." *Biological Conservation* 73: 23–26.

Lott, D. F., and S. C. Minta. 1983a. "Home Ranges of American Bison Cows on Santa Catalina Island, California." *Journal of Mammalogy* 64: 161–62.

———. 1983b. "Random Individual Association and Social Group Instability in American Bison (*Bison bison*)." *Zeitschrift für Tierpsychologie* 61: 153–72.

Lott, D. F., J. H. Shaw, and C. Stormont. 1987. "Should Public Herds Be Trading Bison to Maintain Diversity in the Gene Pool and/or Prevent Inbreeding Depression?" In *North American Bison Workshop*, edited by J. Malcom, 59–60. Missoula, Mont.: U.S. Fish and Wildlife Service and Glacier Natural History Association.

MacEwan, G. 1995. *Buffalo Sacred and Sacrificed*. Edmonton: Alberta Sport, Recreation, Parks, and Wildlife Foundation.

Maher, C. R. 2000. "Quantitative Variation in Ecological and Hormonal Variables Correlates with Spatial Organization of Pronghorn (*Antilocapra americana*) Males." *Behavioral Ecology and Sociobiology* 47: 327–38.

Maher, C. R., and J. A. Byers. 1987. "Age-Related Changes in Reproductive Effort of Male Bison." *Behavioral Ecology and Sociobiology* 21: 91–96.

Manning, R. 1995. *Grassland*. New York: Penguin Books.

Martin, P. S., and R. G. Klein, eds. 1984. *Quaternary Extinctions: A Prehistoric Revolution*. Tucson: University of Arizona Press.

Martin, P. S., and C. R. Szuter. 1999. "War Zones and Game Sinks in Lewis and Clark's West." *Conservation Biology* 13: 36–45.

Matthews, A. 1992. *Where the Buffalo Roam*. New York: Grove Weidenfeld.

Mattson, D. J. 1997. "Use of Ungulates by Yellowstone Grizzly Bears *Ursus arctos*." *Biological Conservation* 81: 161–77.

Mayer, F. H., and C. B. Roth. 1958. *The Buffalo Harvest*. Denver: Sage Books.

McDonald, J. N. 1981. *North American Bison: Their Classification and Evolution*. Berkeley: University of California Press.

McHugh, T. 1972. *The Time of the Buffalo*. New York: Alfred A. Knopf.

Meagher, M., and M. E. Meyer. 1994. "On the Origin of Brucellosis in Bison of Yellowstone National Park: A Review." *Conservation Biology* 8: 645–53.

Mech, L. D. 1991. *The Way of the Wolf*. Stillwater, Minn.: Voyageur Press.

Miller, B. J., and S. H. Anderson. 1993. "Ethology of the Endangered Black-Footed Ferret (*Mustela nigripes*)." *Advances in Ethology* 31: 1–46.

Minta, S. C. 1993. "Sexual Differences in Spatio-temporal Interaction among Badgers." *Oecologia* 96: 402–9.

Minta, S. C., K. A. Minta, and D. F. Lott. 1992. "Hunting Associations between Badgers (*Taxidea taxus*) and Coyotes (*Canis latrans*)." *Journal of Mammalogy* 73: 814–20.

Moehlman, P. D. 1986. "Ecology of Cooperation in Canids." In *Ecological Aspects of Social Evolution: Birds and Mammals*, edited by D. I. Rubenstein and R. W. Wrangham, 64–86. Princeton: Princeton University Press.

Mooring, M. S., and B. L. Hart. 1997. "Self Grooming in Impala Mothers and Lambs: Testing the Body Size and Tick Challenges Principles." *Animal Behaviour* 53: 925–34.

Mooring, M. S., and W. M. Samuel. 1998. "Tick Defense Strategies in Bison: The Role of Grooming and Hair Coat." *Behaviour* 135: 693–719.

Morgan, R. G. 1998. "The Destruction of the Northern Bison Herds." In *International Symposium on Bison Ecology and Management in North America*, edited by L. Irby and J. Knight, 312–25. Bozeman: Montana State University.

Orwell, G. 1950. *Shooting an Elephant, and Other Essays*. New York: Harcourt Brace.

Owen-Smith, R. N. 1988. *Megaherbivores: The Influence of Very Large Body Size on Ecology*. Cambridge: Cambridge University Press.

Owings, D. H., and W. J. Loughry. 1985. "Variation in Snake-Elicited Jump-Yipping by Black-Tailed Prairie Dogs: Ontogeny and Snake-Specificity." *Zeitschrift für Tierpsychologie* 70: 177–200.

Owings, D. H., and E. S. Morton. 1998. *Animal Vocal Communication: A New Approach*. Cambridge: Cambridge University Press.

Parkman, F. 1849. *The California and Oregon Trail: Being Sketches of Prairie and Rocky Mountain Life*. New York: Putnam.

Peden, D. G. 1976. "Botanical Composition of Bison Diets on Shortgrass Plains." *American Midland Naturalist* 96: 225–29.

Price, E. O. 1999. "Behavioral Development in Animals Undergoing Domestication." *Applied Animal Behaviour Science* 65: 245–71.

Pyke, G. H. 1979. "The Economics of Territory Size and Time Budgets in the Golden-Winged Sunbird." *American Naturalist* 114: 131–45.

Robinson, C. M. I. 1995. *The Buffalo Hunters*. Austin, Tex.: State House Press.

Roe, F. G. 1970. *The North American Buffalo: A Critical Study of the Species in Its Wild State*. 2nd ed. Toronto: University of Toronto Press.

Rowe, M. P., R. G. Coss, and D. H. Owings. 1986. "Rattlesnake Rattles and Burrowing Owl Hisses: A Case of Batesian Mimicry." *Ethology* 72: 53–71.

Rubenstein, D. I. 1981. "Behavioural Ecology of Island Feral Horses." *Equine Veterinary Journal* 13: 27–34.

Rutberg, A. T. 1983. "Factors Influencing Dominance Status in American Bison Cows (*Bison bison*)." *Zeitschrift für Tierpsychologie* 63: 206–12.

———. 1986. "Dominance and Its Fitness Consequences in American Bison Cows." *Behaviour* 96: 62–91.

Schaller, G. B. 1972. *The Serengeti Lion: A Study of Predator-Prey Relations.* Chicago: University of Chicago Press.

Seton, E. T. 1910. *Life-Histories of Northern Animals: An Account of the Mammals of Manitoba.* London: Constable.

———. 1927. *Lives of Game Animals.* 4 vols. New York: Doubleday, Doran.

Shaw, J. H. 1993. "American Bison: A Case Study in Conservation Genetics." In *Proceeding: North American Public Bison Herds Symposium,* 3–14, edited by R. E. Walker. N.p.

———. 1995. "How Many Bison Originally Populated Western Rangelands?" *Rangelands* 17 (5): 148–50.

Shaw, J. H., and Tracy S. Carter. 1989. "Calving Patterns among American Bison." *Journal of Wildlife Management* 53: 896–98.

———. 1990. "Bison Movements in Relation to Fire and Seasonality." *Wildlife Society Bulletin* 18: 426–30.

Shaw, J. H., and M. Lee. 1997. "Relative Abundance of Bison, Elk, and Pronghorn on the Southern Plains, 1806–1857." *Plains Anthropologist* 42: 163–72.

Sonenshine, D. E. 1991–93. *Biology of Ticks.* 2 vols. New York: Oxford University Press.

Stegner, W. 1954. *Beyond the Hundredth Meridian: John Wesley Powell and the Second Opening of the West.* Boston: Houghton Mifflin.

Swaisgood, R. R., D. H. Owings, and M. P. Rowe. 1999. "Conflict and Assessment in a Predator-Prey System: Ground Squirrels versus Rattlesnakes." *Animal Behaviour* 57: 1033–44.

Swaisgood, R. R., M. P. Rowe, and D. H. Owings. 1999. "Assessment of Rattlesnake Dangerousness by California Ground Squirrels: Exploitation of Cues from Rattling Sounds." *Animal Behaviour* 57: 1301–10.

van Soest, P. J. 1994. *Nutritional Ecology of the Ruminant.* 2nd ed. Ithaca, N.Y.: Comstock Press.

van Vuren, D. 1987. "Bison West of the Rocky Mountains [USA]: An Alternative Explanation." *Northwest Science* 61: 65–69.

van Zyll de Jong, C. G., C. Gates, H. Reynolds, and W. Olson. 1995. "Phenotypic Variation in Remnant Populations of North American Bison." *Journal of Mammalogy* 76: 391–405.

West, E. 1995. *The Way to the West: Essays on the Central Plains.* Albuquerque: University of New Mexico Press.

West, M. J., A. P. King, and T. M. Freeberg. 1997. "Building a Social Agenda for the Study of Bird Song." In *Social Influences on Vocal Development,* edited by

C. T. Snowdon and M. Hausberger, 41–56. Cambridge: Cambridge University Press.

Whittle, C. L., R. T. Bowyer, T. P. Clausen, and L. K. Duffy. 2000. "Putative Pheromones in Urine of Rutting Male Moose (*Alces alces*): Evolution of Honest Advertisement?" *Journal of Chemical Ecology* 26: 2747–62.

Wilson, G. A., and C. Strobeck. 1999. "Genetic Variation within and Relatedness among Wood and Plains Bison Populations." *Genome* 42: 483–96.

Wolff, J. O. 1998. "Breeding Strategies, Mate Choice, and Reproductive Success in American Bison." *Oikos* 83: 529–44.

Wright, M. 1992. "Le Bois de Vache II: This Chip's for You Too." *Buffalo*, edited by J. E. Foster, D. Harrison, and I. S. MacLaren, 225–24. Edmonton: University of Alberta Press.

Wyckoff, D. G., and W. W. Dalquest. 1997. "From Whence They Came: The Paleontology of Southern Plains Bison." *Plains Anthropologist* 42: 5–32.

Index

Aborted fetuses, 110–11, 113
African antelope, 21, 117, 208n
African buffalo, 21
Aggression: anticipation of, 7–8; artificial
    selection against, 154–56; bellowing stage
    of, 9; by breeding bulls, 5–7, 205n; by
    dominant cows, 23, 24–26, 205nn; by non-
    tending bulls, 8–9; by older bulls, 13–14;
    by pregnant cows, 34; submission to,
    11–13; threatening postures of, 9–11; uri-
    nating/wallowing stage of, 9, 23–24
Alaska: Siberia's connection to, 62; white
    bison of, 54
Allard, Charles, 186, 187
American Bison Society, 183, 187–88, 189
American cheetah, 100
American Fur Company, 173
American lion, 64, 100
American Society for the Prevention of
    Cruelty to Animals (ASPCA), 181
Anaerobic bacteria, 49–50
Anions, Douglas, 75, 99, 207nn, 208n
Anthrax (Bacillus anthracis),
    114–15, 208n
Arctic wolves, 196, 211n
Arkansas River, 83–84, 172
Armitage, Kenneth, 200, 211n
Asian buffalo, 21
ASPCA (American Society for the Prevention
    of Cruelty to Animals), 181
Astor, John Jacob, 173
Atlantic salmon, 199
Attacks. See Aggression
Aurochs, 197

Bacillus anthracis (anthrax), 114–15, 208n
Bacteria: anthrax type of, 114–15, 208n; bru-
    cellosis infection from, 109–11, 113–14,
    208n; colonization/propagation by, 108–9,
    208n; containment strategies for, 111–13
Bacterial digestion of cellulose, 48–50

Badgers: coyotes teaming with, 136–37, 209n;
    digging ability of, 133; social organization
    of, 134, 209n; territoriality of, 135
Badlands National Park, 194, 195, 211n
Bang, Bernhard, 109
Bang's disease. See Brucella abortus
Beaver pelts, 173
Beaver trappers, 180–81
Bellowing of bulls, 9, 17–18
Berger, Joel, 20, 193, 194, 205n
Beringia crossing, 62
Bering Sea, 62
Bialowieza Forest (Poland), 67
Big bluestem grass, 83, 86–87
Big Fifty Sharps rifles, 164, 165–66, 177
Bighorn sheep, 151–52, 181, 210n
Big Medicine (a buffalo), 53–54
Birth of calves, 29–30, 33, 205n, 206n
Birth rate of bison, 26
Bison (Bison bison): anthrax colonization of,
    114–15, 208n; anti-tick strategies of, 116–17,
    208n; athleticism of, 41–46, 206nn; brucel-
    losis infection of, 109–11, 208n; of buffalo
    commons, 92–93, 207n; cowbirds' relation-
    ship with, 105–6, 153; disappearance of,
    170–71, 210nn; diverse attitudes toward,
    182–84; domestication of, 92–93, 195–96,
    197–200; energy efficiencies of, 54–55,
    55–56; European relatives of, 66–67; evolu-
    tion of, 61–66, 206n; fire used to hunt, 85,
    170, 191, 207n; gene pool of, 192–95, 211n;
    grassland's reciprocity with, 86–88, 128;
    grizzly's consumption of, 142–43; human
    encounters with, 151, 153, 156–57, 210nn;
    human taming of, 155–56; hunters' under-
    standing of, 158–61; last hunts for, 140, 179;
    managed grazing of, 191–92; as nonterritor-
    ial, 121; private ownership of, 186, 187; re-
    treat strategy of, 65–66; ruminant digestion
    by, 48–50, 206n; subspecies of, 67–68, 206n;
    terms "bison"/"buffalo," xiv–xv; with

223

Text: 10/14 Palatino
Display: Akzidenz Grotesk Extended; Perpetua
Compositor: Impressions Book and Journal Services, Inc.
Maps: Bill Nelson
Index: Patricia Deminna
Printer and binder: Sheridan Books, Inc.